SpringerBriefs in Complexity

Springer Complexity

Springer Complexity is an interdisciplinary program publishing the best research and academic-level teaching on both fundamental and applied aspects of complex systems—cutting across all traditional disciplines of the natural and life sciences, engineering, economics, medicine, neuroscience, social and computer science.

Complex Systems are systems that comprise many interacting parts with the ability to generate a new quality of macroscopic collective behavior the manifestations of which are the spontaneous formation of distinctive temporal, spatial or functional structures. Models of such systems can be successfully mapped onto quite diverse "real-life" situations like the climate, the coherent emission of light from lasers, chemical reaction-diffusion systems, biological cellular networks, the dynamics of stock markets and of the internet, earthquake statistics and prediction, freeway traffic, the human brain, or the formation of opinions in social systems, to name just some of the popular applications.

Although their scope and methodologies overlap somewhat, one can distinguish the following main concepts and tools: self-organization, nonlinear dynamics, synergetics, turbulence, dynamical systems, catastrophes, instabilities, stochastic processes, chaos, graphs and networks, cellular automata, adaptive systems, genetic algorithms and computational intelligence.

The three major book publication platforms of the Springer Complexity programare the monograph series "Understanding Complex Systems" focusing on the various applications of complexity, the "Springer Series in Synergetics", which is devoted to the quantitative theoretical and methodological foundations, and the "SpringerBriefs in Complexity" which are concise and topical working reports, case-studies, surveys, essays and lecture notes of relevance to the field. In addition to the books in these two core series, the program also incorporates individual titles ranging from textbooks to major reference works.

More information about this series at http://www.springer.com/series/8907

Emanuele Cozzo • Guilherme Ferraz de Arruda
Francisco Aparecido Rodrigues • Yamir Moreno

Multiplex Networks

Basic Formalism and Structural Properties

 Springer

Emanuele Cozzo
Institute for Biocomputation
and Physics of Complex Systems
University of Zaragoza
Zaragoza, Spain

Francisco Aparecido Rodrigues
Departamento de Matemática
Aplicada e Estatística
Universidade de São Paulo
São Carlos, São Paulo, Brazil

Guilherme Ferraz de Arruda
University de São Paulo
São Carlos, São Paulo, Brazil

Yamir Moreno
Institute for Biocomputation
and Physics of Complex Systems (BIFI)
University of Zaragoza
Zaragoza, Spain

ISSN 2191-5326   ISSN 2191-5334   (electronic)
SpringerBriefs in Complexity
ISBN 978-3-319-92254-6   ISBN 978-3-319-92255-3   (eBook)
https://doi.org/10.1007/978-3-319-92255-3

Library of Congress Control Number: 2018943401

Printed on acid-free paper

This Springer imprint is published by the registered company Springer International Publishing AG part of Springer Nature.
The registered company address is: Gewerbestrasse 11, 6330 Cham, Switzerland

Contents

Chapter 1
Introduction

The concept of network served for a long time as a metaphor supporting a *structural approach*, i.e., an approach that puts the accent on the relations among the constituents of a given system, in a wide range of scientific fields, from biology to sociology, and spanning organizational levels from subcellular organization to social organization. In all those fields we can observe a shift from the use of the concept of network as a metaphor to a more substantial notion [12, 43, 82], which has led to what is now known as complex networks science.

The science of complex networks provides an interdisciplinary viewpoint for the study of complex systems, as it constitutes a unifying language that permits to abstract from the specific details of a system to focus on its structure of interactions. The result of this operation of abstraction is a graph model of the system. On its turn, a graph is a specific mathematical object, and a plethora of mathematical tools has been developed to deal with it. Admittedly, the real power of representing a complex system through a graph lies in the hypothesis that the structure and function of the system under study are intimately related to one another. Paraphrasing Wellman [82]: *It is a comprehensive paradigmatic way of taking structure seriously by studying directly how patterns of ties determine the functioning of a system.*

Now, we understand a *complex network* as a system whose patterns of interactions cannot be described by a random graph model with a Poissonian distribution for the connections. From the point of view of a physicist, complex networks are systems that display a strong disorder with large fluctuations of the structural characteristics. As such, the tools developed in condensed matter theory and in statistical physics revealed to be well suited to study the architecture of complex networks [28]. While statistical physics provides complex networks science with a set of tools, graph theory stands at its basis providing a formal language. The first step in complex network research is to represent the structure of the system

© The Author(s) 2018
E. Cozzo et al., *Multiplex Networks*, SpringerBriefs in Complexity,
https://doi.org/10.1007/978-3-319-92255-3_1

under study as a graph, followed by an analysis of the topological features of the obtained representation through a set of informative measures. This first step can be understood as the *formal representation* of the system, while the second one can be seen as the *topological characterization* of the system's structure [22].

Both the peculiar nature of complex networks as topological structures (in comparison, for example, with a lattice) and the particular nature of the system under study push the need for the definition of structural metrics. The *degree distribution* is the most simple example of a structural metric needed for a gross characterization of the inhomogeneity of a networked system. The degree of a node in a network is the number of connections it has to other nodes and the degree distribution is the probability distribution of these degrees over the whole network. While it is really uninformative in a homogeneous system (a lattice, a random regular network, or a poisson random graph), it gives a basic understanding of the degree of the disorder in the case of complex networks. So much so, that the discovery that many networked system has power law degree distributions set the start of the current interest in complex networks science.

On the other hand, the needs of each particular field of research serve as a guide for the definition of particular structural metrics that quantify some relational concepts developed in that field. This is the case of the plethora of centrality measures defined to capture the relation of power in social network analysis [30]. The topological characterization of a complex network also implicitly allows for classifications, either of the constituents of the system, or when comparing different systems. However, *understanding* the structure of a complex network means, roughly speaking, to comprehend what is informative, and what is the result of chance (or of structural constraints). Therefore, a third step in complex network investigation is its statistical characterization, that is, the quantification of the statistical significance of network properties and their dependencies. Crucial in this step is the generation of appropriate null models since once a structural property is recognized as statistically significant, a mechanism for the emergence of such a property could be proposed and further investigated [59].

Finally, the core hypothesis that the structure and the function are intimately related to one another returns. The fourth step is then the functional characterization, i.e., the study of the relations between the structure and the dynamics (as a proxy of the function). From a physicist point of view, the interest is in studying the critical properties of dynamical processes defined on complex networks, as models of real processes and emergent functions. The crucial point is that many of the peculiar critical effects showed by processes defined on complex networks are closely related and universal for different models, basically reinforcing the hypothesis of the relation between the structure and the function, together with the "statistical physical" approach [28]. On the other hand, when a particular system is under study, this part of the investigation deals with the task of finding the structural properties that may explain the observed phenomena.

1.1   From Simple Networks to Multiplex Networks

The concept of multiplex networks may be anchored in communication media or in the multiplicity of roles and milieux. When focusing on the former aspect, one realizes that the constituents of a complex system continuously switch among a variety of media to make the system perform properly. On the other hand, focusing on the latter, one takes into account the fact that interactions are always context dependent as well as integrated through different contexts.

The term *multiplexity* was coined in early 1962 in the social anthropology framework by Max Gluckman (in [32]) to denote "the coexistence of different normative elements in a social relationship," i.e., the coexistence of distinct roles in a social relationship. While this first definition focus on context and roles, Kapferer offered a second definition based on the overlap of different activities and/or exchanges in relationships, focusing on the social relationship as a medium for the exchanges of different types of information [40]. The duality between media and roles in founding the multiplexity of social relations is still present in the contemporary debate, with authors like David Bolter and Richard Gusin [7] advocating the former, and others like Lee Rainie and Barry Wellman the latter [64]. However, whether defined by roles or media, multiplexity always refers to "multiple bases for interaction" in a network [78].

It is indubitable that new push for the formal and quantitative research in multiplex networks comes from the social and technological revolution brought by the Internet and mobile connections. Chats, online social networks, and a plethora of other human-to-human machine mediated channels of communications, together with the possibility of being always online (hyper-connectivity [81]), have accelerated the proliferation of layers that makes "the sociality." Although it has a longer history in the field of social sciences, the concept of multiplexity, and consequently of multiplex networks, is not restricted to them. For example, it is gaining an important role in contemporary Biology, where we can observe the same shift from its use as a metaphor to a more substantial notion of the concept of multiplex networks. In particular, it is associated with the method of integration of multiple sets of omic data (data from genomics, proteomics, and metabolomics) on the same population; as well as to the case of meta-genomic networks where the dynamical interactions between the genome of the host and that of the microbes living in it, the cross-talk being mediated by chemical and ecological interactions. As with the case of social multiplex networks, also in biology the origin of the renovated interest in multiplex networks is largely due to a technological jump that has made it possible the availability of large and diverse amounts of data coming from very different experimental settings.

Also in the traditional field of transportation networks, the notions of multiplexity and multiplex networks have a natural translation in different modes of transportation connecting the same physical location in a city, a country, or on the globe. Finally, in the field of engineering and critical infrastructures, the concept of multiplexity applies to the interdependence of different lifelines [14].

Even if the notion of multiplexity was introduced years ago, the discussions included few analytical tools to accompany them. This situation arose for a simple reason: although many aspects of single-layer networks are well understood, it is challenging to properly generalize even the simplest concept to multiplex networks. Theoretical developments on multilayer networks in general, and on multiplex networks in particular, have gained steam only in the last few years, for different reasons, among which surely stands the technological revolution represented by the digitalization and the social transformations that have accompanied it, as we have mentioned before.

Now, we understand multiplex networks as a nonlinear superposition of complex networks, where components, being them social actors, genes, devices, or physical locations, interact through a variety of different relationships and communication channels, which we conceptualize as different layers of the multiplex network. This conceptualization of a multiplex network poses a number of challenges to the science of complex networks: from the formal description of a complex network (starting from the fact that a constituent of a networked system, represented by a node in a traditional single layer network, is no more an "elementary" unit, indeed it has an internal—possibly complex—structure that must be formally represented), to the ultimate goal of understanding how this new level of structural complexity represented by the interaction of different layers reveals itself in the functioning of the system.

In this Springer Brief, we provide, based on our own research experience, a formal introduction to the subject of Multiplex Networks. Despite the relative youth of this topic, there are already many results available in the literature, both concerning the structural and topological characterization of these networks as well as on the dynamics on top of them. Additionally, ours is not the only approach to characterize multiplex networks, and therefore, the reader might also find several references in which the notation and terminology are different to the one discussed here. In this sense, ours is a "biassed" introduction, although we think it is a natural one and easier to follow for those that already have some basic knowledge of single layer networks. It is also worth stressing that we do not revise all the body of works that deal with the topology of multiplex networks, but those to which we, as researcher, have contributed the most. Topics such as community detection, time dependent multiplex networks, and the definition and use of several (perhaps more secondary) metrics have not been addressed here. A final limitation of the present book is that it only presents some aspects of general dynamical processes (mainly diffusion and spreading dynamics) whenever we need them to illustrate the need to use the multiplex perspective.

The main reason to proceed in this way is because we believe that a first introduction to the topic needs to address what topological aspects are relevant, and then discuss what is their impact on dynamics (a second argument is that there are simply too many dynamics that have been studied already and discussing them is beyond the scope of the present text). The monograph is organized in a way that once the basic definitions are understood, the reader can navigate through the different chapters independently (most of the time). We hope that both the choice of topics and the way they are presented help the interested reader to introduce herself in the subject as well as that the present contribution can be used for an introductory course to multilayer networks at the postgraduate level.

Acknowledgements We would like to thank all those who have collaborated with us during the last several years in our journey exploring multilayer networks. The work presented here has benefited from the following funders: CNPq (grant 305940/2010-4), Fapesp (Grants 2013/26416-9, 2012/25219-2, and 2015/07463-1), the FPI program of the Government of Aragón, Spain, the Government of Aragón, Spain through a grant to the group FENOL, MINECO, and FEDER funds (grant FIS2014-55867-P), and the European Commission FET Projects Multiplex (grant 317532) and Plexmath (grant 317614).

Chapter 2
Multiplex Networks: Basic Definition and Formalism

In this chapter, we present and define multiplex networks as they will be used in this book. It is common to introduce multiplex networks as a particular specification of the more general notion of multilayer networks [44], conversely, we prefer to have the former as a primary object. We then show how this structure can be represented by adjacency matrices, introducing the notion of "supra-adjacency" matrix. A different algebraic representation of multiplex networks is possible by means of higher-order tensors, the supra-adjacency matrix being a particular flattening of an adjacency tensor representing the multiplex network under some assumptions [25]. We will briefly introduce this formalism in Chap. 7 and show some applications.

This introductory chapter represents an effort to set a formal language in this area, and it is intended to be general and complete enough as to deal with the most diverse cases. Although it might seem pedantic, setting a rigorous algebraic formalism is crucial to make it possible and, in a certain sense, automatic, furthermore complex reasoning, as well as to design data structures and algorithms.

2.1 Graph Representation

A networked system $\mathbf{N}$ is naturally represented by a graph. A graph is a tuple $G(V, E)$, where V is a set of nodes, and $E \subseteq V \times V$ is a set of edges that connects a pair of nodes. Nodes represent the components of the system, while edges represent interactions or relations among them. If an edge exists in G between node u and node v, i.e., $(u, v) \in E$, they are said to be adjacent, and we indicate the adjacency relation with the symbol $\sim$, i.e., we will write $u \sim v$ if $(u, v) \in E$. When needed, we write $u \overset{G}{\sim} v$ to explicitly state that the adjacency relation is referred to the particular graph G.

© The Author(s) 2018
E. Cozzo et al., *Multiplex Networks*, SpringerBriefs in Complexity,
https://doi.org/10.1007/978-3-319-92255-3_2

In order to represent a networked system in which different types of relations or interactions exist between the components—a multiplex network—the notion of layer must be introduced. Let $L = \{1, \ldots, m\}$ be an index set, which we call the *layer set*. A layer is an index that represents a particular type of interaction or relation. $\mid L \mid = m$ is the number of layers in the multiplex network, i.e., the number of different kind of interactions/relations in the system. Now, consider a set of nodes V, where nodes represent the components of the system, and let $G_P = (V, L, P)$ be a binary relation, where $P \subseteq V \times L$. The statement $(u, \alpha) \in P$, with $u \in V$, and $\alpha \in L$, is read *node u participates in layer α*. We call the ordered pair $(u, \alpha) \in P$ a *node-layer pair* and we say that the node-layer pair (u, α) is the representative of node u in layer α, thus P is the set of the node-layer pairs. In other words, we are attaching etiquettes to nodes that specify in which type of relations (layers) the considered node participates in.

$G_P = (V, L, P)$ can be interpreted as a (bipartite) graph where P is the edge set. $\mid P \mid = N$ is the number of node-layer pairs, while $\mid V \mid = n$ is the number of nodes. If each node $u \in V$ has a representative in each layer, i.e., $P = V \times L$, we call the multiplex a *node-aligned multiplex*, and we have that $\mid P \mid = mn$. As we shall see later, things are always simpler when the multiplex is node-aligned.

In this way, each system of relations or interactions of different kinds is naturally represented by a graph $G_\beta(V_\beta, E_\beta)$, where $V_\beta = \{(u, \alpha) \in P \mid \alpha = \beta\}$, that is, V_β is a subset of P composed by all the node-layer pairs that have the particular index β as second element. In other words, it is the set of all the representatives of the node set in a particular layer. The edge set $E_\beta \subseteq V_\beta \times V_\beta$ represents interactions or relations of a particular type between the components of the systems. We call $G_\beta(V_\beta, E_\beta)$ a *layer-graph* and we can consider the set of all layer-graphs $M = \{G_\alpha\}_{\alpha \in L}$. $\mid V_\beta \mid = n_\beta$ is the number of node-layer pairs in layer β. For node-aligned multiplex networks we have $n_\alpha = n$, $\forall \alpha \in L$.

Finally, consider the graph G_C on P in which there is an edge between two node-layer pairs (u, α) and (v, β) if and only if $u = v$; that is, when the two edges in the graph G_P are incident on the same node $u \in V$, which means that the two node-layer pairs represent the same node in different layers. We call $G_C(P, E_C)$ the coupling graph. It is easy to realize that the coupling graph is formed by $n = \mid P \mid$ disconnected components that are either complete graphs or isolated nodes. Each component is formed by all the representatives of a node in different layers, and we call the components of G_C *supra-nodes*.

We are now in the position to say that a multiplex network is represented by the quadruple $\mathcal{M} = (V, L, P, M)$:

- the node set V represents the components of the system,
- the layer set L represents different types of relations or interactions in the system,
- the participation graph G_P encodes the information about what node takes part in a particular type of relation and defines the representative of each component in each type of relation, i.e., the node-layer pair,

Fig. 2.1 The multiplex
network is represented by the
quadruple
$\mathcal{M} = (V, L, P, M)$:, i.e., the
layer set $L = \{a_1, a_2\}$, the set
of the node-layer pairs $P =$
$\{(1, a_1), (2, a_1), (3, a_1), (2, a_2), (3, a_2)\}$,
the node set $V = \{1, 2, 3\}$.
Supra-nodes are $[(1, a_1)]$,
$[(2, a_1), (2, a_2)]$, and
$[(3, a_1), (3, a_2)]$

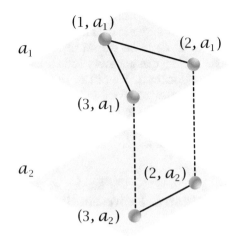

- the layer-graphs M represent the networks of interactions of a particular type
  between the components, i.e., the networks of representatives of the components
  of the system.

Figure 2.1 illustrates the notation used to describe the multilayer organization.

Next, consider the union of all the layer-graphs, i.e., $G_l = \bigcup_\alpha G_\alpha$. We call such
a graph the *intra-layer graph*. Note that, if each layer-graph is connected, this graph
is formed by m disconnected components, one for each layer-graph.

Finally, we can define the graph $G_\mathcal{M} = G_l \cup G_C$, which we call the *supra-graph*.
$G_\mathcal{M}$ is a synthetic representation of a multiplex network. Note that supra-nodes are
cliques[1] in $G_\mathcal{M}$.

To summarize, up to now, we have two different entities representing the components of a multiplex network: nodes and node-layer pairs. A node corresponds to a
"physical object," while node-layer pairs are different instances of the same object.
For instance a node could represent an online user, while node-layer pairs would
represent different accounts of the same user in different online social networks;
or a node could represent a social actor, while node-layer pairs would represent
different social roles (friend, worker, family member) of the same social actor; or a
node could stand for a location in a transportation network, while node-layer pairs
would represent stations of different transportation modes (e.g., streets, highways,
and subways).

The connection between nodes and node-layer pairs is given by the notion
of supra-nodes: i.e., cliques in the supra-graph formed by node-layer pairs that
are instances of the same object. Moreover, for clarity, we denote nodes using
the symbols u, v, w; for brevity, we may indicate a node-layer pair with a single

---

[1]A clique, C, in an undirected graph $G = (V, E)$ is a subset of the vertices, $C \subseteq V$, such that
every two distinct vertices are adjacent.

symbol instead of using the ordered pair (u, α), and we will use the symbols i, j, h. To round off the basic definitions used henceforth, let's also define $l(u) = \{(u, \alpha) \in P \mid \alpha \in L\}$ to be the set of all node-layer pairs that correspond to the same node u. Note that not every node has a representative in every layer, and $l(u)$ may have cardinality 1. We call $\kappa_u = |l(u)|$ the *multiplexity degree* of the node u, that is, the number of layers in which an instance of the same object u appears. We also define $l^{-1}(i)$ to be the unique node that corresponds to the node-layer pair i. Furthermore, when it is clear from the context, we may refer to node-layer pairs simply as nodes.

2.2   Matrix Representation

Given a graph $G(V, E)$, we can associate to it a matrix $\mathbf{A}(G)$ whose elements $a_{uv} = 1_{\underset{u \sim v}{G}}$, where 1_x is the indicator function, i.e., it is equal to one if the x is true, otherwise it is zero. The matrix $\mathbf{A}(G)$ is called the adjacency matrix of G, and by identifying a network $\mathbf{N}$ with its graph representation, we say that $\mathbf{A}(G)$ is the adjacency matrix of $\mathbf{N}$.

We can consider the adjacency matrix of each of the graphs introduced in the previous section. The adjacency matrix of a layer graph G_α is a $n_\alpha \times n_\alpha$ symmetric matrix $\mathbf{A}^{(\alpha)} = \mathbf{A}(G_\alpha)$, with $a_{ij}^\alpha = 1_{\underset{i \sim j}{G^\alpha}}$, i.e., if and only if there is an edge between i and j in G^α. We call them *layer adjacency matrices*.

Likewise, the adjacency matrix of G_P is an $n \times m$ matrix $\mathbf{P} = \mathbf{P}(G_P)$, with $p_{u\alpha} = 1_{\underset{u \sim \alpha}{G_P}}$, i.e., if and only if there is an edge between the node u and the layer α in the participation graph; that is, only if node u participates in layer α. We call it the *participation matrix*. The adjacency matrix of the coupling graph G_C is an $N \times N$ matrix $\mathcal{C} = \mathbf{C}(G_C)$, with $c_{ij} = 1_{\underset{i \sim j}{G_C}}$, i.e., if and only if there is an edge between node-layer pair i and j in G_C; that is, if they are representatives of the same node in different layers. We can arrange the rows and the columns of $\mathcal{C}$ such that node-layer pairs of the same layer are contiguous. It results that $\mathcal{C}$ is a block matrix with zero diagonal blocks. Besides, rows and columns can be arranged in a way such that the off-diagonal blocks are diagonals. Thus, $c_{ij} = 1$, with $i, j = 1, \ldots, N$ represents an edge between a node-layer pair in layer 1 and a node-layer pair in layer 2 if $i < n_1$ and $n_1 < j < n_2$. We call this the *standard labeling* and we assume that node-layer pairs are always labeled this way. Note that this labeling also induces a labeling of node-layer pairs in layer-graphs such that the same row and column in different layer adjacency matrices correspond to the representative of the same node in different layers.

In addition, when in the previous section we said that we may use a single symbol i instead of using the ordered pair (u, α) to indicate a node-layer pair, we were stating that we identify a node layer pair with its index in the corresponding layer adjacency matrix. In the same way, $l(u)$ can now be interpreted as the set of layer adjacency matrix indexes that correspond to a given node u.

Considering the example in Fig. 2.1, we have:

$$\mathbf{A}^{(a_1)} = \begin{bmatrix} 0 & 1 & 1 \\ 1 & 0 & 0 \\ 1 & 0 & 0 \end{bmatrix},$$

$$\mathbf{A}^{(a_2)} = \begin{bmatrix} 0 & 1 \\ 1 & 0 \end{bmatrix},$$

and

$$\mathbf{P} = \begin{bmatrix} 1 & 0 \\ 1 & 1 \\ 1 & 1 \end{bmatrix}.$$

Note that in this example the labeling is not standard. One has to exchange the labeling of nodes 1 and 3 in order to have a standard labeling.

2.2.1 The Supra-Adjacency Matrix

Given a supra-graph $G_{\mathcal{M}}$, we consider its adjacency matrix $\mathbf{A}(G_{\mathcal{M}})$ and we call it the *supra-adjacency* matrix $\bar{A}$. Just as $G_{\mathcal{M}}$, $\bar{A}$ is a synthetic representation of the whole multiplex $\mathcal{M}$. By definition, assuming the standard labeling, it can be obtained from the intra-layer adjacency matrices and the coupling matrix in the following way:

$$\bar{A} = \bigoplus_{\alpha} \mathbf{A}^{\alpha} + \mathcal{C}, \tag{2.1}$$

where $\bigoplus$ represents the direct sum. We also define $\mathcal{A} = \bigoplus_{\alpha} \mathbf{A}^{\alpha}$, and we call it the *intra-layer adjacency* matrix. By definition, the intra-layer adjacency matrix is the adjacency matrix of the intra-layer graph G_l, $\mathcal{A} = \mathbf{A}(G_l)$. Figure 2.2 shows the supra-adjacency matrix, the intra-layer adjacency matrix, and the coupling matrix of a multiplex network.

$\bar{A}$ takes a very simple form in the case of node-aligned multiplex networks, that is

$$\bar{A} = \mathcal{A} + \mathbf{K}_m \otimes \mathbf{I}_n, \tag{2.2}$$

where $\otimes$ represents the Kronecker product, $\mathbf{K}_m$ is the adjacency matrix of a complete graph on m nodes, and I_n is the $n \times n$ identity matrix.

Fig. 2.2 Example of a
multiplex network. The
structure of each layer is
represented by an adjacency
matrix $\mathcal{A}^{(i)}$, where $i = 1, 2$.
$\mathcal{C}_{(lm)}$ stores the connections
between layers l and m. Note
that the number of nodes in
each layer is not the same

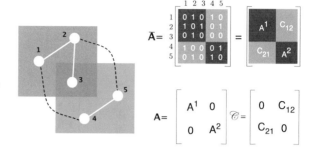

It is even simpler when layer-graphs are identical:

$$\bar{A} = \mathbf{I}_m \otimes \mathbf{A} + \mathbf{K}_m \otimes \mathbf{I}_n, \tag{2.3}$$

where $\mathbf{A}$ is the adjacency matrix of each identical layer graph. Equation (2.3) is just
the Kronecker sum of $\mathbf{A}$ and $\mathbf{K}_m$.

For instance, if we have two layers, $L = \{1, 2\}$ with edge set $E_1 = \{[(1, 1), (2, 1)], [(2, 1), (3, 1)]\}$ and $\{E_2 = [(1, 2), (2, 2)], [(2, 2), (3, 2)]\}$, we
have that

$$\bar{A} = \mathbf{I}_2 \otimes \mathbf{A} + \mathbf{K}_2 \otimes \mathbf{I}_3$$

$$= \begin{bmatrix} 1 & 0 \\ 0 & 1 \end{bmatrix} \otimes \begin{bmatrix} 0 & 1 & 0 \\ 1 & 0 & 1 \\ 0 & 1 & 0 \end{bmatrix} + \begin{bmatrix} 0 & 1 \\ 1 & 0 \end{bmatrix} \otimes \begin{bmatrix} 1 & 0 & 0 \\ 0 & 1 & 0 \\ 0 & 0 & 1 \end{bmatrix}$$

$$= \begin{bmatrix} \mathbf{A} & 0 \\ 0 & \mathbf{A} \end{bmatrix} + \begin{bmatrix} 0 & \mathbf{I}_3 \\ \mathbf{I}_3 & 0 \end{bmatrix} = \begin{bmatrix} \mathbf{A} & \mathbf{I}_3 \\ \mathbf{I}_3 & \mathbf{A} \end{bmatrix}$$

Some basic metrics are easily calculated from the supra-adjacency matrix. The
degree of a node-layer i is the number of node-layers connected to it by an edge in
$G_{\mathcal{M}}$ and is given by

$$K_i = \sum_j \bar{A}_{ij}. \tag{2.4}$$

Sometimes we write $i(\alpha)$ as an index, instead of simply i, to explicitly indicate that
the node-layer i is in layer α even if the index i already uniquely indicates a node-
layer pair. Since $\bar{A}$ can be read as a block matrix, with the $\mathbf{A}^{(\alpha)}$ on the diagonal
blocks, the index $i(\alpha)$ can be interpreted as block index. It is also useful to define
the following quantities

$$e_\alpha = \sum_{\beta < \alpha} n_\beta, \tag{2.5}$$

which we call the excess index of layer α. Notice that n_β is the number of node-layer pairs in layer β. The layer-degree of a node-layer i, $k_{i(\alpha)}$, is the number of neighbors it has in G^α, i.e., $k_{i(\alpha)} = \sum_j a_{ij}^\alpha$. By definition of $\bar{A}$

$$k_{i(\alpha)} = \sum_{j=1+e_\alpha}^{n_\alpha+e_\alpha} \bar{A}_{ij}. \tag{2.6}$$

The coupling degree of a node-layer i, $c_{i(\alpha)}$, is the number of neighbors it has in the coupling graph, i.e., $c_{i(\alpha)} = \sum_j c_{ij}$. From $\bar{A}$ we get

$$c_{i(\alpha)} = \sum_{\substack{j<e_\alpha, \\ j>n_\alpha+e_\alpha}} \bar{A}_{ij}. \tag{2.7}$$

By definition

$$c_i = \kappa_{l^{-1}(i)} - 1. \tag{2.8}$$

Finally, we note that the degree of a node-layer can be expressed as

$$K_{i(\alpha)} = \sum_j \bar{A}_{ij} = k_{i(\alpha)} + c_{i(\alpha)}. \tag{2.9}$$

Equation (2.9) explicitly expresses the fact that the degree of a node-layer pair is the sum of its layer-degree plus its coupling-degree.

2.2.2 The Supra-Laplacian Matrix

Generally, the Laplacian matrix, or simply the Laplacian, of a graph with adjacency matrix $\mathbf{A}(G)$ is given by

$$\mathbf{L}(G) = \mathbf{D} - \mathbf{A}(G) \tag{2.10}$$

where $\mathbf{D}(G) = diag(k_1, k_2, \dots)$ is the degree matrix.

Thus, it is natural to define the *supra-Laplacian* matrix of a multiplex network as the Laplacian of its supra-graph

$$\bar{\mathcal{L}} = \bar{\mathcal{D}} - \bar{A}, \tag{2.11}$$

where $\bar{\mathcal{D}} = diag(K_1, K_2, \dots, K_N)$ is the degree matrix.

Besides, we can define the layer-Laplacian of each layer-graph G_α as

$$\mathbf{L}^{(\alpha)} = \mathbf{D}^{(\alpha)} - \mathbf{A}^{(\alpha)}, \tag{2.12}$$

and the Laplacian of the coupling graph

$$\mathcal{L}_C = \Delta - \mathcal{C} \tag{2.13}$$

where $\Delta = diag(c_1, c_2, \ldots, c_N)$ is the coupling-degree matrix.

By definition, we have

$$\bar{\mathcal{L}} = \bigoplus_\alpha \mathbf{L}^{(\alpha)} + \mathcal{L}_C. \tag{2.14}$$

As it was the case of the supra-adjacency matrix, Eq. (2.14) takes a very simple form in the case of a node-aligned multiplex, i.e.,

$$\bar{\mathcal{L}} = \bigoplus_\alpha (\mathbf{L}^{(\alpha)} + (m-1)I_N) - \mathbf{K}_m \otimes I_n, \tag{2.15}$$

and when layer-graphs are identical:

$$\bar{\mathcal{L}} = \mathbf{I}_m \otimes (\mathbf{L} + (m-1)I_n) - \mathbf{K}_m \otimes I_n, \tag{2.16}$$

where $\mathbf{L}$ is the Laplacian of each identical layer-graph.

2.2.3 Multiplex Walk Matrices

A walk on a graph is defined as a sequence of adjacent vertices. The length of a walk is the number of edges it contains. For a simple graph (which has no multiple edges), a walk may be specified completely by an ordered list of vertices [83]. A step is the elementary component of a walk, i.e., two adjacent nodes.

Here, supra-walk is defined as a walk on a multiplex network in which, either before or after each intra-layer step, a walk can either continue on the same layer or change to an adjacent layer. We represent this choice by the matrix:

$$\widehat{\mathcal{C}}(\beta, \gamma) = \beta \mathcal{I} + \gamma \mathcal{C} \tag{2.17}$$

where $\mathcal{I}$ is the $N \times N$ identity matrix, the parameter β is a weight that accounts for the walk staying in the current layer, and γ is a weight that accounts for the walk stepping to another layer. In a supra-walk, a supra-step consists either of only a single intra-layer step or of a step that includes both an intra-layer step changing from one layer to another (either before or after having an intra-layer step). In the latter type of supra-step, note that we are disallowing two consecutive inter-layer steps. In other words, supra-walks are walks on the supra-graph $G_\mathcal{M}$ with this latter prescription.

Roughly speaking, a *multiplex walk matrix* is a matrix that encodes the permissible steps in a multiplex network. The matrix $\mathcal{A}\widehat{\mathcal{C}}$ encodes the steps in which after each intra-layer step a walk can continue on the same layer. On the other hand, the matrix $\widehat{\mathcal{C}}\mathcal{A}$ encodes the steps in which before each intra-layer step a walk can continue on the same layer. Both matrices $\mathcal{A}\widehat{\mathcal{C}}$ and $\widehat{\mathcal{C}}\mathcal{A}$ can be interpreted as the adjacency matrix of a directed (possibly weighted) graph. We call such graphs *auxiliary supra-graph*.

In general, depending on the rules prescribed to walk the multiplex, one can define an auxiliary supra-graph G_M whose adjacency matrix is $\mathcal{M} = \mathcal{M}(\mathcal{A}, \mathcal{C})$. It should be noted that, by definition, the supra-adjacency matrix is also a walk matrix. The need of such matrices comes from the fact that, as we will see in the next chapter, often it is of interest to treat intra and inter-layer edges differently, where changing layer is an action of a different nature with respect to going from a node to a different one.

2.3   Coarse-Graining Representation of a Multiplex Network

Because of the structure of a multiplex network, it is natural to try to aggregate the interaction pattern of each different layer in a single network somehow. An operation that is called *dimensionality reduction*, whereas the result of such operation leads to an object named *aggregate network*. Several candidates for the aggregate network have been proposed in the literature such as the average network [73], the overlapping network [6], the projected monoplex network or the overlay network [25]. As in [69], we claim that the natural definition of an aggregate network is given by the notion of quotient network. In addition, the quotient network framework allows to introduce in a symmetric way another aggregate network, the *network of layers*, that encodes the connectivity pattern between layers. In a sense that will be more clear in Chap. 4, the notion of quotient graph underpins the notion of multiplex network. Figure 2.3 shows, in a schematic way, a multilayer network and the two quotient networks that can be derived from it. We next describe in detail how they are obtained.

2.3.1   Mathematical Background

Let us first provide a brief, but self-contained description of network quotients.

2.3.1.1   Adjacency and Laplacian Matrices

Suppose that $\{V_1, \ldots, V_m\}$ is a partition of the node set of a graph $G(V, E)$ with adjacency matrix $\mathbf{A}(G)$, and write $n_i = |V_i|$.

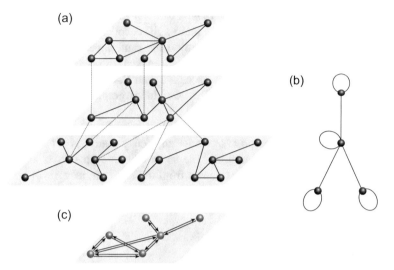

Fig. 2.3 Schematic representation of a multiplex network with 4 layers and 8 nodes per layer (**a**), and its two quotients: the network of layers (**b**), and the aggregate network (**c**). In (**a**), dashed lines represent inter-layer edges. The quotient (**b**) is undirected, as all layers have the same number of nodes. The quotient (**c**) is only partially drawn, it is directed, and the edge thickness is proportional to the weight. The network of layers (**b**) corresponds to the layer interconnection structure, while the aggregate network (**c**) represents the superposition of all the layers onto one. In this sense, they can be thought of as "horizontal" and "vertical" quotients, as the figure suggests. Both quotients clearly represent a dimensionality reduction or coarsening of the original multilayer network

The *quotient graph* $\mathcal{Q}$ of G is a coarsening of G with respect to the partition. It has one node per cluster V_i, and an edge from V_i to V_j weighted by an average connectivity from V_i to V_j

$$b_{ij} = \frac{1}{\sigma} \sum_{\substack{k \in V_i \\ l \in V_j}} a_{kl}, \tag{2.18}$$

where we have a choice for the size parameter σ: we will use either $\sigma_i = n_i$, or $\sigma_j = n_j$, or $\sigma_{ij} = \sqrt{n_i}\sqrt{n_j}$. We call the corresponding graph the *left quotient*, the *right quotient*, and the *symmetric quotient*, respectively. Fortunately, the matrix $\mathbf{B} = (b_{ij})$ has the same eigenvalues for the three choices of σ (see below). We refer by *quotient graph* to any of these three spectrally equivalent graphs with adjacency matrix $\mathbf{B}$. Observe that the symmetric quotient is undirected, while the left and right quotients are not, unless all clusters have the same size, $n_i = n_j$ for all i, j.

The quotient formalism holds in more generality for any real symmetric matrix, as we explain here. Let $\mathbf{A} = (a_{ij})$ be any real symmetric $n \times n$ matrix. Write $X = \{1, 2, \ldots, n\}$, let $\{X_1, \ldots, X_m\}$ be a partition of X, and let $n_i = |X_i|$. We write A_{ij} for the submatrix consisting of the intersection of the k-rows and l-columns of A such that $k \in X_i$ and $l \in X_j$. In particular, A_{ij} is an $n_i \times n_j$ matrix. Define b_{ij} as the average row sum of A_{ij},

$$b_{ij} = \frac{1}{n_i} \sum_{\substack{k \in X_i \\ l \in X_j}} a_{kl}. \tag{2.19}$$

The $m \times m$ matrix $Q_l(\mathbf{A}) = (b_{ij})$ is called the *left quotient matrix of* $\mathbf{A}$ with respect to the partition $\{X_1, \ldots, X_m\}$. We can express $Q_l(\mathbf{A})$ in matrix form, as follows. Let $\mathbf{S} = (s_{ij})$ be the $n \times m$ *characteristic matrix* of the partition, that is, $s_{ij} = 1$ if $i \in X_j$, and 0 otherwise. Then $\mathbf{S}^T \mathbf{A} \mathbf{S}$ is the matrix of coefficient sums of the submatrices A_{ij}, and, hence, $Q_l(\mathbf{A}) = \Lambda^{-1} \mathbf{S}^T A \mathbf{S}$, where $\Lambda = \mathrm{diag}(n_1, \ldots, n_m)$.

There are two alternatives to $Q_l(\mathbf{A})$, called the *right quotient* and the *symmetric quotient*, written $Q_r(\mathbf{A})$ and $Q_s(\mathbf{A})$. They correspond to replacing $1/n_i$ in (2.19) by $1/n_j$ respectively $1/\sqrt{n_i}\sqrt{n_j}$. In matrix form, we have $Q_r(\mathbf{A}) = \mathbf{S}^T \mathbf{A} \mathbf{S} \Lambda^{-1}$ and $Q_s(\mathbf{A}) = \Lambda^{-1/2} \mathbf{S}^T \mathbf{A} \mathbf{S} \Lambda^{-1/2}$. Note that $Q_l(\mathbf{A})$ is the transpose of $Q_r(\mathbf{A})$, and they are not symmetric unless $n_i = n_j$ for all i, j.

Nevertheless, these three matrices have the same spectrum (the proof is straight-forward):

Lemma 1 *Let* $\mathbf{X}, \mathbf{D}$ *be* $m \times m$ *matrices, with* $\mathbf{D}$ *diagonal. Then the matrices* $\mathbf{DX}$, $\mathbf{XD}$ *and* $\mathbf{D}^{1/2}\mathbf{XD}^{1/2}$ *have all the same spectrum.*

Summarizing, the *left quotient*, the *right quotient*, and the *symmetric quotient* graph of a graph G with adjacency matrix $\mathbf{A}$ is the graph Q with adjacency matrix $\mathbf{B} = Q_l(\mathbf{A})$, $\mathbf{B} = Q_r(\mathbf{A})$, and $\mathbf{B} = Q_s(\mathbf{A})$, respectively. Consider the left quotient of $\mathbf{A}$ with respect to the partition. Observe that the row sums of $Q_l(\mathbf{A})$ are

$$\bar{d}_i = \frac{1}{n_i} \sum_{k \in V_i} d_k, \tag{2.20}$$

the average node degrees in V_i. Let $\overline{\mathbf{D}}$ be the diagonal matrix of the average node degrees. Then we define the *quotient Laplacian* as the matrix

$$L_Q = \overline{D} - Q_l(\mathbf{A}). \tag{2.21}$$

(See Chap. 4 for a full discussion on this choice.) Moreover, let $\tilde{Q}$ be the *loopless quotient* of G, that is, the quotient network Q with all the self-loops removed. As the quotient Laplacian ignores self-loops (see Chap. 4), we have $L_Q = L_{\tilde{Q}}$.

2.3.1.2 Regular Quotients

A partition $\{V_1, \ldots, V_m\}$ of the node set is called *equitable* if the number of edges (taking weights into account) from a node in V_i to any node in V_j is independent of the chosen node in V_i

$$\sum_{l \in V_j} a_{kl} = \sum_{l \in V_j} a_{k'l} \quad \text{for all } k, k' \in V_i, \tag{2.22}$$

for all i, j. This indicates a regularity condition on the connection pattern between (and within) clusters. If the partition is equitable, we call the quotient network *regular*. A source of regular quotients is network symmetries [50, 51]. We call a partition *almost equitable* if the condition (2.22) is satisfied for all $i \neq j$ (but not necessarily for $i = j$), that is, if the regularity condition is satisfied after ignoring the intra-cluster edges. In this case, we call the quotient graph $\mathcal{Q}$ *almost regular*. Note that the quotient $\mathcal{Q}$ being almost regular is equivalent to the loop-less quotient $\tilde{\mathcal{Q}}$ being regular.

2.3.2 The Aggregate Network

Define the node characteristic matrix $\mathbf{S}_n = (s_{iu})$. $\mathbf{S}_n$ is an $N \times n$ matrix with $s_{iu} = 1$ if and only if the node-layer i is a representative of node u, i.e., it is in the connected component u of the graph G_C. We call it a characteristic matrix since nodes partition the node-layer set and $\mathbf{S}_n$ is the characteristic matrix of that partition.

Then, the adjacency matrix of the aggregate network is given by:

$$\tilde{\mathbf{A}} = \Lambda^{-1} \mathbf{S}_n^T \bar{\mathcal{A}} \mathbf{S}_n, \tag{2.23}$$

where $\Lambda = diag\{\kappa_1, \ldots, \kappa_n\}$ is the multiplexity degree matrix.

We also define the *average connectivity* between nodes u and v as

$$d_{uv} = \frac{1}{\kappa_u} \sum_{\substack{i \in l(u) \\ j \in l(v)}} \bar{\mathcal{A}}_{ij}, \tag{2.24}$$

and write d_u for d_{uu}. In this way, in an aggregate network, each node has a self-loop weighted by d_u, and a directed edge from u to v weighted by d_{uv}. Note that in general the aggregate network is directed. However, if the multiplex network is node-aligned, then the aggregate network is not directed.

We also define a loop-less aggregate network, that is just the aggregate network without self-loops, i.e.,

$$\tilde{\mathbf{W}} = \tilde{\mathcal{A}} - diag(\tilde{\mathcal{A}}) \tag{2.25}$$

It is worth noting that

$$\tilde{\mathbf{W}} = \Lambda^{-1} \mathbf{S}_n^T \mathcal{A} \mathbf{S}_n. \tag{2.26}$$

Finally, we define the sum aggregate network as

$$\mathbf{W} = \mathbf{S}_n^T \mathcal{A} \mathbf{S}_n, \tag{2.27}$$

and note that for node-aligned multiplex networks we have

$$\mathbf{W} = m\tilde{\mathbf{W}}. \tag{2.28}$$

2.3.3   The Network of Layers

Likewise, define the layer characteristic matrix $\mathbf{S}_l = \{s_{i\alpha}\}$ as an $N \times m$ matrix with $s_{i\alpha} = 1$ only if the node-layer i is in layer α, i.e., in the connected component α of the graph G_l. We call it a characteristic matrix since it is the characteristic matrix of the partition of the node-layer set induced by layers.

In the same way, the network of layers has adjacency matrix given by

$$\tilde{\mathbf{A}}_l = \Lambda^{-1}\mathbf{S}_l^T \bar{\mathcal{A}}\mathbf{S}_l, \tag{2.29}$$

where $\Lambda^{-1} = diag\{n_1, \ldots, n_m\}$.

Finally, we define the *average inter-layer degree* from α to β as

$$d^{\alpha\beta} = \frac{1}{n_\alpha} \sum_{\substack{i \in V_\alpha \\ j \in V_\beta}} a_{ij}. \tag{2.30}$$

This represents the average connectivity from a node in G_α to any node in G_β. If $\alpha = \beta$ we write d^α for $d^{\alpha\alpha}$, and call it the *average intra-layer degree*. Thus, each node corresponds to a layer, with a self-loop weighted by the average intra-layer degree d^α, and there is a directed edge from layer α to layer β weighted by the average inter-layer degree $d^{\alpha\beta}$.

2.4   Supra-Walk Matrices and Loopless Aggregate Network

By definition, the quantity $(\mathcal{A}\widehat{\mathcal{C}})_{ij}^n + (\widehat{\mathcal{C}}\mathcal{A})_{ij}^n$ counts the number (the weight) of different supra-walks (or cycles if $i = j$) of length l between node-layer pairs i and j. From the symmetric properties of $\mathcal{A}$ and $\mathcal{C}$ it follows that

$$(\widehat{\mathcal{C}}\mathcal{A})_{ij}^l = (\mathcal{A}\widehat{\mathcal{C}})_{ji}^l. \tag{2.31}$$

Thus, the number (the weight) of supra-walks of length n between node-layer pairs i and j is given by:

$$\mathcal{N}(l)_{ij} = (\mathcal{A}\widehat{\mathcal{C}})_{ij}^l + (\mathcal{A}\widehat{\mathcal{C}}^l)_{ij}^T. \tag{2.32}$$

Here, we want to give the relation between the number of supra-walks in the multiplex network and the number of walks in the sum aggregate network when changing layer has no cost. We have the following result connecting the walk matrices $\mathcal{A}\widehat{\mathcal{C}}(\beta, \gamma)$ and $\widehat{\mathcal{C}}(\beta, \gamma)$ and the sum aggregate network in the particular cases in which changing layer has no cost:

Lemma 2 *The right (left) quotient of* $\mathcal{A}\widehat{\mathcal{C}}(\beta, \gamma)$ *(* $\widehat{\mathcal{C}}(\beta, \gamma)\mathcal{A}$ *) is equal to the sum-aggregate network when* $\beta = \gamma$

Proof Observe that $\widehat{C}(\beta, \beta) = \beta \mathbf{S}\mathbf{S}^T$ and remember that $\mathbf{S}^T\mathbf{S} = \Lambda$, then $Q_R(\mathcal{A}\widehat{\mathcal{C}}) = \mathbf{S}^T \mathcal{A}.\beta \mathbf{S}\mathbf{S}^T \mathbf{S}\Lambda^{-1} = \beta \mathbf{S}^T \mathcal{A}\mathbf{S} = \mathbf{W}$ The result for $(\widehat{\mathcal{C}}\mathcal{A})$ follows by transposition.   □

Besides, we can prove the following

Lemma 3 $Q_L(\widehat{\mathcal{C}}(\beta, \beta)\mathcal{A})\mathbf{S}^T = \mathbf{S}^T \mathcal{A}\widehat{\mathcal{C}}(\beta, \beta)$

Proof $Q_L(\widehat{\mathcal{C}}(\beta, \beta)\mathcal{A})\mathbf{S}^T = \beta \Lambda^{-1}\mathbf{S}^T \widehat{\mathcal{C}}\mathcal{A}\mathbf{S}\mathbf{S}^T = \alpha \Lambda^{-1}\mathbf{S}^T \mathbf{S}\mathbf{S}^T \mathcal{A}\mathbf{S}\mathbf{S}^T = \alpha \mathbf{S}^T \mathcal{A}\mathbf{S}\mathbf{S}^T = \mathbf{S}^T \mathcal{A}\widehat{\mathcal{C}}(\beta, \beta)$
The result for $\widehat{\mathcal{C}}(\beta, \beta)\mathcal{A}$ follows by transposition.   □

The previous result follows from the fact that if a walk exists from i to j in $\mathcal{A}\widehat{\mathcal{C}}$, then it exists a walk from j to i in $\widehat{\mathcal{C}}\mathcal{A}$ and vice versa. It is interesting to note that this implies that the in (out) degree of a node layer i in $\mathcal{A}\widehat{\mathcal{C}}$ $(\widehat{\mathcal{C}}\mathcal{A})$ with respect to an element of the partition only depends on the element of the partition it belongs to. Armed with Lemmas 2 and 3 we can prove that

Theorem 1 $\sum_{i \in l(u), j \in l(v)} \mathcal{N}(l)_{ij} = 2\mathbf{W}^l_{uv}$

The proof follows from Lemma 3 by induction. The relation established in this section will be crucial in the next chapter in order to correctly generalize different structural metrics defined for single-layer networks to multiplex networks.

Chapter 3
Structural Metrics

A structural metric of a network is a measure of some property directly dependent on the system of relations between the components of the network, i.e., by representing the network with a graph, a structural metric is a measure of a property that depends on the edge set. Since there is a correspondence between graph and adjacency matrix, a structural metric can be expressed as a function of the adjacency matrix, but it is not necessary. This fact differentiates structural metrics and other kinds of metrics, such as spectral metrics, that are defined only once an adjacency matrix is introduced.

Structural metrics can be local or global. A local metric p measures the property of a single node or pair of nodes, and we refer to the value of that metric on a node i or a pair of nodes i, j as p_i, p_{ij} respectively. The global version P of p measures the corresponding overall property of the network. In general, a global metric is defined as a mean of local ones. An example of a structural metric is the connectivity k. As we have seen, the connectivity k_i of a node i is the number of neighbors the node i has. The global connectivity is defined as the mean connectivity $K = \frac{1}{N} \sum_i k_i$. The characteristic path length L is a global metric defined as $L = \frac{1}{n(n-1)} \sum_{i \neq j} l_{ij}$, where l_{ij} is the geodesic distance between the nodes i and j measured as the minimum number of edges connecting i and j.

Although in the two examples given above the global metric is simply the mean of the local one, it is not always the case, as for the clustering coefficient (see note 2 in Sect. 3.1). The term local and global may have a different meaning in this context; in fact, they may refer to the topological scale at which the system is considered. In this sense, the connectivity is a local measure since it takes into account only the first neighbors of a node, while the geodesic distance between two nodes is a global metric since it takes into account the whole network. The quantitative description of structural network properties is a core task of complex network research. Firstly, it allows for the classification of different structures and the description of different categories. On the other hand, it is the first step for the investigation of the relations between structure and dynamic/function. Finally, it

© The Author(s) 2018
E. Cozzo et al., *Multiplex Networks*, SpringerBriefs in Complexity,
https://doi.org/10.1007/978-3-319-92255-3_3

allows the construction of models that reproduce the structural features of an empiric system under study, as well as to inquire if a property of a system is the result of chance or if it reveals something on the particular way the system evolved. An example of the latter is the fact that recognizing that the clustering coefficient of an empiric social network is on average greater than that of a random graph allowed to propose the *triadic closure* as a crucial mechanism in the evolution of social networks. The other way around, the quantitative evaluation of the clustering coefficient between model networks and empirical ones allows the validation of the *triadic closure* hypothesis. Thus, it is crucial to define a set of structural metrics for multiplex networks.

Here, we dare to suggest a list of requirements a structural metric should fulfill in order to be properly defined. A structural metric for multiplex networks should

- reduce to the ordinary single-layer metric (if defined) when layers reduce to one,
- be defined for node-layer pairs,
- be defined for non-node-aligned multiplex networks.

The first requirement refers to the generalization of standard single-layer metrics to multiplex networks. It seems reasonable, although it is not trivial. In fact, usually generalizing "the naive way" leads to metrics that do not fulfill this requirement. We discuss this point in the next session in the particular case of the clustering coefficient. The second requirement takes into account the fact that node-layer pairs are the basic objects that build up a multiplex network, thus, in general, to define a metric only on some version of the aggregate network is not enough. The third requirement comes from the fact that, although it is easier to deal with node-aligned multiplex networks from an analytical point of view, real world multiplex networks in general are not node-aligned. Because of that, it is worth defining metrics for the general case, even when an analytic treatment is only possible in the node-aligned case.

An additional requirement is needed only in the case of intensive metrics:

- For a multiplex of identical layers when changing layer has no cost, an intensive structural metric should take the same value when measured on the multiplex network and on one layer taken as an isolated network.

This last requirement, that asks for a sort of "normalization," is needed in order to avoid spurious amplification of the value that a quantity takes just because of the number of layers. This *list of requirements* has not the pretension to be interpreted as a set of axioms and surely it is not definitive nor complete, but in our opinion it has the power to guide the generalization of standard single-layer metrics to multiplex networks in a systematic way, as well as to guide the theoretical development of new genuinely multiplex metrics.

In summary, we can recognize that it is insufficient to generalize existing diagnostics in a naïve manner and that one must instead construct their generalizations from first principles. In the following sections of this chapter, we will build on the basic notion of walks and cycles to properly generalize clustering coefficients and subgraphs centrality to multiplex networks.

3.1 Structure of Triadic Relations in Multiplex Networks

In the present section, we focus on triadic relations, which are used to describe the simplest and most fundamental type of transitivity in networks [42, 49, 55, 79, 80]. Following [19], we present multiplex generalizations of clustering coefficients, which can be done in myriad ways, and (as we will illustrate) the most appropriate generalization depends on the application under study.

There have been several attempts to define multiplex clustering coefficients [5, 6, 9, 10, 21], but there are significant shortcomings in these definitions. For example, some of them do not reduce to the standard single-layer clustering coefficient or are not properly normalized [19]. The fact that some definitions of multiplex clustering coefficients are mostly *ad hoc* makes them difficult to interpret. We present the definitions given in [19] that build on the basic concepts of walks and cycles to obtain a transparent and general definition of transitivity. This approach also guarantees that clustering coefficients are always properly normalized. It reduces to a weighted clustering coefficient [85] of an aggregated network for particular values of the parameters; this allows comparison with existing single-layer diagnostics. Two additional, very important issues, are also addressed: (1) Multiplex networks have many types of connections, and the given definition of multiplex clustering coefficients are (by construction) decomposable, so that the contribution of each type of connection is explicit; (2) because the given notion of multiplex clustering coefficients builds on walks and cycles, it does not require every node to have a representative in all layers, which removes a major (and very unrealistic) simplification that is used in other definitions.

In an unweighted single-layer network, the local clustering coefficient C_u of node u is the number of triangles (i.e., triads) that includes node u divided by the number of connected triples with node u in the center [55, 80]. The local clustering coefficient is a measure of transitivity [49], and it can be interpreted as the density of a focal node's neighborhood. For the present purposes, it is convenient to define the local clustering coefficient C_u as the number of 3-cycles t_u that starts and ends at the focal node u divided by the number of 3-cycles d_u such that the second step of the cycle occurs in a complete graph (i.e., assuming that the neighborhood of the focal node is as dense as possible).[1] In mathematical terms, $t_u = (\mathbf{A}^3)_{uu}$ and $d_u = (\mathbf{AFA})_{uu}$, where $\mathbf{A}$ is the adjacency matrix of the graph and $\mathbf{F}$ is the adjacency matrix of a complete graph with no self-edges. (In other words, $\mathbf{F} = \mathbf{J} - \mathbf{I}$, where $\mathbf{J}$ is a complete square matrix of 1s and $\mathbf{I}$ is the identity matrix.)

The *local clustering coefficient* is thus given by the formulas $C_u = t_u/d_u$. This is equivalent to the usual definition of the local clustering coefficient: $C_u = t_u/(k_u(k_u - 1))$, where $k_u \geq 2$ is the degree of node u (the local clustering coefficient being 0 for nodes of degree 0 and 1). One can calculate a single global

---

[1]Note that we use the term "cycle" to refer to a walk that starts and ends at the same physical node u. It is permissible (and relevant) to return to the same node via a different layer from the one that was used originally to leave the node.

clustering coefficient for a monoplex network either by averaging C_u over all nodes or by computing $C = \frac{\sum_u t_u}{\sum_u d_u}$. Henceforth, we will use the term *global clustering coefficient* for the latter quantity.[2] In the following, we will give the definitions of the clustering coefficients for multiplex networks.

3.1.1 Triads on Multiplex Networks

In addition to 3-cycles (i.e., triads) that occur within a single layer, multiplex networks also contain cycles that can traverse different additional layers but still have 3 intra-layer steps, thus involving three nodes. Such cycles are important for the analysis of transitivity in multiplex networks. In social networks, for example, transitivity involves social ties across multiple social environments [76, 79]. In transportation networks, there typically exist several means of transportation to return to one's starting location, and different combinations of transportation modes are important in different cities [31]. For dynamical processes on multiplex networks, it is important to consider 3-cycles that traverse different numbers of layers, so one needs to take them into account when defining a multiplex clustering coefficient. For these reasons, it is crucial to build the clustering coefficient on the notion of supra-walk. Thus, the number of 3-cycles for node-layer pair i is then

$$t_{M,i} = [(\mathcal{A}\widehat{\mathcal{C}})^3 + (\widehat{\mathcal{C}}\mathcal{A})^3]_{ii} , \tag{3.1}$$

where the first term corresponds to cycles in which the inter-layer step is taken after an intra-layer one and the second term corresponds to cycles in which the inter-layer step is taken before an intra-layer one, see Chap. 2. The subscript M refers to the particular way that we define a supra-walk in a multiplex network through the multiplex walk matrices $\mathcal{A}\widehat{\mathcal{C}}$ and $\widehat{\mathcal{C}}\mathcal{A}$. However, one can also use other types of supra-walks, and we will use different subscripts when we refer to them. Exploiting the fact that both $\mathcal{A}$ and $\widehat{\mathcal{C}}$ are symmetric, Eq. (3.1) becomes

$$t_{M,i} = 2[(\mathcal{A}\widehat{\mathcal{C}})^3]_{ii} . \tag{3.2}$$

It is useful to decompose multiplex clustering coefficients that are defined in terms of multilayer cycles into the so-called *elementary cycles* by expanding Eq. (3.2) and writing it in terms of the matrices $\mathcal{A}$ and $\mathcal{C}$. That is, write $t_{M,i} = \sum_{\mathcal{E} \in \mathscr{E}} w_{\mathcal{E}}(\mathcal{E})_{ii}$, where $\mathscr{E}$ denotes the set of elementary cycles and $w_{\mathcal{E}}$ are weights of different elementary cycles. One can use symmetries in the definition of cycles

---

[2] The definition we adopt for the global clustering coefficient is an example of a global structural metric that is not defined as the mean value over all the nodes of its local version. Actually, it is defined as the ratio between the mean number of closed triples and the mean number of open triples.

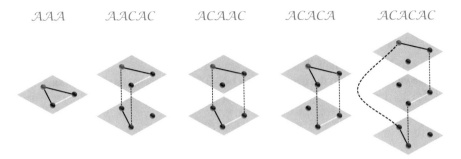

$\mathcal{A}\mathcal{A}\mathcal{A}$ $\quad$ $\mathcal{A}\mathcal{A}C\mathcal{A}C$ $\quad$ $\mathcal{A}C\mathcal{A}\mathcal{A}C$ $\quad$ $\mathcal{A}C\mathcal{A}C\mathcal{A}$ $\quad$ $\mathcal{A}C\mathcal{A}C\mathcal{A}C$

Fig. 3.1 Sketch of the elementary cycles $\mathcal{A}\mathcal{A}\mathcal{A}$, $\mathcal{A}\mathcal{A}C\mathcal{A}C$, $\mathcal{A}C\mathcal{A}\mathcal{A}C$, $\mathcal{A}C\mathcal{A}C\mathcal{A}$, and $\mathcal{A}C\mathcal{A}C\mathcal{A}C$. The orange node is the starting point of the cycle. The intra-layer edges are the solid lines, and the intra-layer edges are the dotted curves. In each case, the yellow line represents the second intra-layer step

and thereby express all of the elementary cycles in a standard form with terms from the set $\mathcal{E} = \{\mathcal{A}\mathcal{A}\mathcal{A}, \mathcal{A}\mathcal{A}C\mathcal{A}C, \mathcal{A}C\mathcal{A}\mathcal{A}C, \mathcal{A}C\mathcal{A}C\mathcal{A}, \mathcal{A}C\mathcal{A}C\mathcal{A}C\}$. See Fig. 3.1 for an illustration of elementary cycles and Sect. 3.1.2 for details on deriving the elementary cycles. Note that some of the alternative definitions of a 3-cycle lead to more elementary cycles than the ones that we just enumerated.

To define multiplex clustering coefficients, both the number $t_{*,i}$ of cycles and a normalization $d_{*,i}$ are needed. The symbol $*$ stands for any type of cycle: the 3-cycle defined above, an elementary cycle, or the alternative definitions of 3-cycles based on alternative ways to walk the multiplex networks, i.e., on different walk matrices.

Choosing a particular definition implies a given way to calculate the associated expression for $t_{*,i}$. To determine the normalization, it is natural to follow the same procedure as with monoplex clustering coefficients and use a complete multiplex network $\mathcal{F} = \bigoplus_\alpha \mathbf{F}^{(\alpha)}$, where $\mathbf{F}^{(\alpha)} = \mathbf{J}^{(\alpha)} - \mathbf{I}^{(\alpha)}$ is the adjacency matrix for a complete graph on layer α. We can then proceed from any definition of $t_{*,i}$ to $d_{*,i}$ by replacing the second intra-layer step with a step in the complete multiplex network. For example, we obtain $d_{M,i} = 2(\widehat{A}\widehat{C}\widehat{F}\widehat{C}\widehat{A}\widehat{C})_{ii}$ for $t_{M,i} = 2[(\widehat{A}\widehat{C})^3]_{ii}$. Similarly, one can use any other definition of a cycle as a starting point for defining a multiplex clustering coefficient.

The definition of local and global clustering coefficients for multiplex networks analogously to single-layer networks follows from the above formulation. We can calculate a natural multiplex analogue to the usual single-layer local clustering coefficient for any node-layer pair i of the multiplex network. Additionally, a node u allows an intermediate description for clustering between local and the global clustering coefficients. The definitions are

$$c_{*,i} = \frac{t_{*,i}}{d_{*,i}}, \qquad (3.3)$$

$$C_{*,u} = \frac{\sum_{i \in l(u)} t_{*,i}}{\sum_{i \in l(u)} d_{*,i}}, \qquad (3.4)$$

$$C_* = \frac{\sum_i t_{*,i}}{\sum_i d_{*,i}}, \tag{3.5}$$

where $l(u)$ is as in Chap. 2.

We can decompose the expression in Eq. (3.5) in terms of the contributions from cycles that traverse exactly one, two, and three layers (i.e., for $m = 1, 2, 3$) to obtain

$$t_{*,i} = t_{*,1,i}\beta^3 + t_{*,2,i}\beta\gamma^2 + t_{*,3,i}\gamma^3, \tag{3.6}$$

$$d_{*,i} = d_{*,1,i}\beta^3 + d_{*,2,i}\beta\gamma^2 + d_{*,3,i}\gamma^3, \tag{3.7}$$

$$C_*^{(m)} = \frac{\sum_i t_{*,m,i}}{\sum_i d_{*,m,i}}. \tag{3.8}$$

We can similarly decompose Eqs. (3.3) and (3.4). Using the decomposition in Eq. (3.6) yields an alternative way to average over contributions from the three types of cycles:

$$C_*(\omega_1, \omega_2, \omega_3) = \sum_m^3 \omega_m C_*^{(m)}, \tag{3.9}$$

where $\vec{\omega}$ is a vector that gives the relative weights of the different contributions. $C_*^{(1)}$, $C_*^{(2)}$, and $C_*^{(3)}$ are said *layer-decomposed clustering coefficients*. There are also analogs of Eq. (3.9) for the clustering coefficients defined in Eqs. (3.3) and (3.4). Each of the clustering coefficients in Eqs. (3.3)–(3.5) depends on the values of the parameters β and γ, but the dependence vanishes if $\beta = \gamma$. Unless we explicitly indicate otherwise, we assume in the following calculations that $\beta = \gamma$.

3.1.2 Expressing Clustering Coefficients Using Elementary 3-Cycles

An elementary cycle is defined as a term that consists of products of the matrices $\mathcal{A}$ and $\mathcal{C}$, i.e., there are no sums, after one expands the expression for a cycle (which is a weighted sum of such terms). Because we are only interested in the diagonal elements of the terms and we consider only undirected layer-graphs and coupling graphs, we can transpose the terms and still write them in terms of the matrices $\mathcal{A}$ and $\mathcal{C}$ rather than also using their transposes. There are also multiple ways of writing nonsymmetric elementary cycles [e.g., $(\mathcal{A}\mathcal{A}\mathcal{C}\mathcal{A}\mathcal{C})_{ii} = (\mathcal{C}\mathcal{A}\mathcal{C}\mathcal{A}\mathcal{A})_{ii}$].

The adopted convention is that in which all elementary cycles are transposed so that it is selected the one in which the first element is $\mathcal{A}$ rather than $\mathcal{C}$ when comparing the two versions of the term from left to right. That is, for two equivalent terms, we choose the one that comes first in alphabetical order. The set of elementary 3-cycles is thus $\mathscr{E} = \{\mathcal{A}\mathcal{A}\mathcal{A}, \mathcal{A}\mathcal{A}\mathcal{C}\mathcal{A}\mathcal{C}, \mathcal{A}\mathcal{C}\mathcal{A}\mathcal{A}\mathcal{C}, \mathcal{A}\mathcal{C}\mathcal{A}\mathcal{C}\mathcal{A}, \mathcal{A}\mathcal{C}\mathcal{A}\mathcal{C}\mathcal{A}\mathcal{C}, \mathcal{C}\mathcal{A}\mathcal{A}\mathcal{A}\mathcal{C}, \mathcal{C}\mathcal{A}\mathcal{A}\mathcal{C}\mathcal{A}\mathcal{C}, \mathcal{C}\mathcal{A}\mathcal{C}\mathcal{A}\mathcal{C}\mathcal{A}\mathcal{C}\}$.

As usual, the normalization formulas can be obtained replacing the second $\mathcal{A}$ term with $\mathcal{F}$. This yields a standard form for any local multiplex clustering coefficients

$$c_{*,i} = \frac{t_{*,i}}{d_{*,i}}, \tag{3.10}$$

with

$$
\begin{aligned}
t_{*,i} = [&w_{\mathcal{AAA}}\mathcal{AAA} + w_{\mathcal{AACAC}}\mathcal{AACAC} + w_{\mathcal{ACAAC}}\mathcal{ACAAC} \\
&+ w_{\mathcal{ACACA}}\mathcal{ACACA} + w_{\mathcal{ACACAC}}\mathcal{ACACAC} \\
&+ w_{\mathcal{CAAAC}}\mathcal{CAAAC} + w_{\mathcal{CAACAC}}\mathcal{CAACAC} \\
&+ w_{\mathcal{CACACAC}}\mathcal{CACACAC}]_{ii}
\end{aligned} \tag{3.11}
$$

$$
\begin{aligned}
d_{*,i} = [&w_{\mathcal{AAA}}\mathcal{AFA} + w_{\mathcal{AACAC}}\mathcal{AFCAC} + w_{\mathcal{ACAAC}}\mathcal{ACFAC} \\
&+ w_{\mathcal{ACACA}}\mathcal{ACFCA} + w_{\mathcal{ACACAC}}\mathcal{ACFCAC} \\
&+ w_{\mathcal{CAAAC}}\mathcal{CAFAC} + w_{\mathcal{CAACAC}}\mathcal{CAFCAC} \\
&+ w_{\mathcal{CACACAC}}\mathcal{CACFCAC}]_{ii},
\end{aligned} \tag{3.12}
$$

where i is a node-layer pair and the $w_{\mathcal{E}}$ coefficients are scalars that correspond to the weights for each type of elementary cycle (these weights are different for different types of clustering coefficients). Note that the parameters β and γ have been absorbed into these coefficients. Possible elementary cycles are shown in Fig. 3.1.

3.1.3 Clustering Coefficients for Aggregated Networks

A common way to study multiplex networks is to aggregate layers to obtain either multi-graphs or weighted networks, where the number of edges or the weight of an edge is the number of different types of edges between a pair of nodes [44]. One can then use any of the numerous ways to define clustering coefficients for weighted single-layer networks [58, 70] to calculate clustering coefficients for the aggregated network. One of the weighted clustering coefficients is a special case of our multiplex clustering coefficient. References [1, 36, 85] calculated a weighted clustering coefficient as

$$C_{Z,u[1]} = \frac{\sum_{vw} W_{uv}W_{vw}W_{wu}}{w_{\max}\sum_{v\neq w} W_{uv}W_{uw}} = \frac{(\mathbf{W}^3)_{uu}}{((\mathbf{W}(w_{\max}\mathbf{F})\mathbf{W})_{uu}}, \tag{3.13}$$

where $\mathbf{W}$ is the sum aggregate adjacency matrix as defined in Chap. 2, the quantity $w_{\max} = \max_{u,v} \mathbf{W}_{uv}$ is the maximum weight in $\mathbf{W}$, and $\mathbf{F}$ is the adjacency matrix of the complete unweighted graph. We can define the global version C_Z of $C_{Z,u}$ by summing over all the nodes in the numerator and the denominator of Eq. (3.13) (analogously to Eq. (3.5)).

For node-aligned multiplex networks, the clustering coefficients $C_{Z,u}$ and C_Z are related to our multiplex clustering coefficients $C_{M,u}$ and C_M. Letting $\beta = \gamma = 1$ and summing over all layers yields $\sum_{i \in l(u)} ((\mathcal{A}\mathcal{C})^3)_{ii} = (\mathbf{W}^3)_{uu}$ (see Sect. 2.4). That is, in this special case, the weighted clustering coefficients $C_{Z,u}$ and C_Z are equivalent to the corresponding multiplex clustering coefficients $C_{M,u}$ and C_M. That is, $C_{M,u}(\beta = \gamma) = w_{\max} C_{Z,u}$ and $C_M(\beta = \gamma) = w_{\max} C_Z$. Note that this relationship between our multiplex clustering coefficient and the weighted clustering coefficient in Eq. (3.13) is only true for node-aligned multiplex networks. If some nodes are not shared among all layers, then the normalization of our multiplex clustering coefficient depends on how many nodes are present in the local neighborhood of the focal node. This contrasts with the "global" normalization by $w_{\max}$ used by the weighted clustering coefficient in Eq. (3.13).

3.1.4 Clustering Coefficients in Erdős-Rényi (ER) Networks

Almost all empirical networks contain some amount of transitivity, and it is often desirable to know if a network contains more transitivity than would be expected by chance. In order to examine this question, one typically compares values of the clustering coefficient of a network to what would be expected from some random network that acts as a null model. The simplest random network to use is an Erdős-Rényi (ER) network. In this section, we give formulas for expected clustering coefficients in node-aligned multiplex networks in which each intra-layer network is an ER network that is created independently of other intra-layer networks and the inter-layer connections are created as described in Chap. 2.

The expected value of the local clustering coefficient in an unweighted monoplex ER network is equal to the probability p of an edge to exist. That is, the density of the neighborhood of a node, measured by the local clustering coefficient, has the same expectation as the density of the entire network for an ensemble of ER networks. In multiplex networks with ER intra-layer graphs with connection probabilities p_α, the same result holds only when all the layers are statistically identical (i.e., $p_\alpha = p$ for all α). Note that this is true even if the network is not node-aligned. However, heterogeneity among layers complicates the behavior of clustering coefficients. If the layers have different connection probabilities, then the expected value of the mean clustering coefficient is a nontrivial function of the connection probabilities. In particular, it is not always equal to the mean of the connection probabilities. For example, the formulas for the expected global layer-decomposed clustering coefficients are

$$\langle C_M^{(1)} \rangle = \frac{\sum_\alpha p_\alpha^3}{\sum_\alpha p_\alpha^2} \equiv \frac{\overline{p^3}}{\overline{p^2}}, \tag{3.14}$$

$$\langle C_M^{(2)} \rangle = \frac{3 \sum_{\alpha \neq \kappa} p_\alpha p_\kappa^2}{(b-1) \sum_\alpha p_\alpha^2 + 2 \sum_{\alpha \neq \kappa} p_\alpha p_\kappa}, \tag{3.15}$$

$$\langle C_M^{(3)} \rangle = \frac{\sum_{\alpha \neq \kappa, \kappa \neq \mu, \mu \neq \alpha} p_\alpha p_\kappa p_\mu}{(b-2) \sum_{\alpha \neq \kappa} p_\alpha p_\kappa}. \tag{3.16}$$

The expected values of the local clustering coefficients in node-aligned ER multiplex networks are

$$\langle c_{A,A,A,i} \rangle = \frac{1}{b} \sum_{\alpha \in L} p_\alpha \equiv \overline{p}, \tag{3.17}$$

$$\langle c_{A,AC,AC,i} \rangle = \frac{1}{b} \sum_{\alpha \in L} p_\alpha \equiv \overline{p}, \tag{3.18}$$

$$\langle c_{AC,A,AC,i} \rangle = \frac{1}{b} \sum_{\alpha \in L} \frac{\sum_{\kappa \neq \alpha} p_\kappa^2}{\sum_{\kappa \neq \alpha} p_\kappa}, \tag{3.19}$$

$$\langle c_{AC,AC,A,i} \rangle = \frac{1}{b} \sum_{\alpha \in L} p_\alpha \equiv \overline{p}, \tag{3.20}$$

$$\langle c_{AC,AC,AC,i} \rangle = \frac{1}{b(b-1)} \sum_{\alpha \in L} \frac{\sum_{\kappa \neq \alpha; \mu \neq \kappa, \alpha} p_\kappa p_\mu}{\sum_{\kappa \neq \alpha} p_\kappa}. \tag{3.21}$$

Note that $c_{M,i}^{(1)} = c_{A,A,A,i}$ and $c_{M,i}^{(3)} = c_{AC,AC,AC,i}$, but the 2-layer clustering coefficient $c_{M,i}^{(2)}$ arises from a weighted sum of contributions from three different elementary cycles.

In Fig. 3.2 it is illustrated the behavior of the global and local clustering coefficients in multiplex networks in which the layers consist of ER networks with varying amounts of heterogeneity in the intra-layer edge densities. Although the global and mean local clustering coefficients are equal to each other when averaged over ensembles of single layer ER networks, the same is not true for multiplex networks with ER layers unless the layers have the same value of the parameter p. The global clustering coefficients give more weight than the mean local clustering coefficients to denser layers. This is evident for the clustering coefficients $c_{M,i}^{(1)}$ and $C_M^{(1)}$, for which the ensemble average of the mean of the local clustering coefficient $c_{M,i}^{(1)}$ is always equal to the mean edge density, whereas the ensemble average of the global clustering coefficient $C_M^{(1)}$ has values that are greater than or equal to the mean edge density. This effect is a good example of a case in which the situation in multiplex networks differs from the results and intuition from single layer networks.

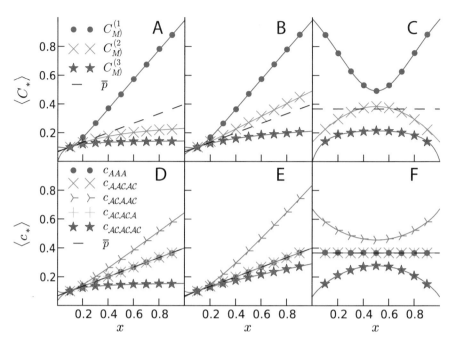

Fig. 3.2 (**a, b, c**) Global and (**d, e, f**) local multiplex clustering coefficients in multiplex networks that consist of ER layers. The markers give the results of simulations of 100-node ER node-aligned multiplex networks that we average over 10 realizations. The solid curves are theoretical approximations (see Eqs. (3.14)–(3.16) of the main text). Panels (**a, c, d, f**) show the results for three-layer networks, and panels (**b, e**) show the results for six-layer networks. The ER edge probabilities of the layers are (**a, d**) $\{0.1, 0.1, x\}$, (**b, e**) $\{0.1, 0.1, 0.1, 0.1, x, x\}$, and (**c, f**) $\{0.1, x, 1 - x\}$

In particular, failing to take into account the heterogeneity of edge densities in multiplex networks can lead to incorrect or misleading results when trying to distinguish among values of a clustering coefficient that are what one would expect from an ER random network versus those that are a signature of a triadic-closure process (see Fig. 3.2).

3.2 Transitivity in Empirical Multiplex Networks

In Table 3.1, we show the values of layer-decomposed global clustering coefficients for multiplex networks (four social networks and two transportation networks) calculated in [19]. Note that the two transportation networks are not "node-aligned." To help give context to the values, the table also includes the clustering-coefficient values obtained for ER networks with matching edge densities in each layer. Those examples show that multiplex clustering coefficients give insights that are impossible to infer by calculating weighted clustering coefficients for aggregated networks or even by calculating them separately for each layer of a multiplex network.

For each social network in Table 3.1, note that $C_M < C_M^{(1)}$ and $C_M^{(1)} > C_M^{(2)} > C_M^{(3)}$. Consequently, the primary contribution to the triadic structure of these multiplex networks arises from 3-cycles that stay within a given layer. We observe that all clustering coefficients exhibit larger inter-layer transitivities than would be expected in ER networks with identical edge densities, and that the same ordering relationship (i.e., $C_M^{(1)} > C_M^{(2)} > C_M^{(3)}$) holds. This observation suggests that triadic-closure mechanisms in social networks cannot be considered purely at the aggregated network level, because these mechanisms appear to be more effective inside of layers than between layers. For example, if there is a connection between individuals u and v and also a connection between v and w in the same layer, then it is more likely that u and w "meet" in the same layer than in some other layer.

The transportation networks examined exhibit the opposite pattern with respect the social networks. For example, for the London Underground ("Tube") network, in which each layer corresponds to a line, $C_M^{(3)} > C_M^{(2)} > C_M^{(1)}$ holds. This reflects the fact that single lines in the Tube are designed to avoid redundant connections. A single-layer triangle would require a line to make a loop among 3 stations. Two-layer triangles, which are a bit more frequent than single-layer ones, entail that two lines run in almost parallel directions and that one line jumps over a single station. For 3-layer triangles, the geographical constraints do not matter because one can construct a triangle with three straight lines.

In Fig. 3.3a, we show a comparison of the layer-decomposed local clustering coefficients. Observe that the condition $c_{M,i}^{(1)} > c_{M,i}^2 > c_{M,i}^{(3)}$ holds for most of the nodes. In Fig. 3.3b, the expected values of the clustering coefficients of nodes in a network generated with the configuration model[3] is subtracted from the corresponding values of the clustering coefficient observed in the data to discern whether we should also expect to observe the relative order of the local clustering coefficients in an associated random network (with the same layer densities and degree sequences as the data). Similar to the results for global clustering coefficients, we see that taking a null model into account lessens—but does not remove—the difference between the coefficients that count different numbers of layers.

It is known that local values of the clustering coefficient are typically correlated with the degree of the nodes in single-layer networks. To compare, Fig. 3.4a shows how the different multiplex clustering coefficients depend on the unweighted degrees of the nodes in the aggregated network for the Kapferer tailor shop. Note that the relative order of the mean clustering coefficients is independent of the degree. In Fig. 3.4b, we illustrate the fact that the aggregated network for the airline transportation network exhibits a nonconstant difference between the curves of

---

[3]We use the configuration model instead of an ER network as a null model because the local clustering coefficient values are typically correlated with nodes degrees in single layer networks [55], and an ER-network null model would not preserve degree sequence.

Table 3.1 Clustering coefficients C_M, $C_M^{(1)}$, $C_M^{(2)}$, and $C_M^{(3)}$ that correspond, respectively, to the global, one-layer, two-layer, and three-layer clustering coefficients for various multiplex networks

		Tailor Shop	Management	Families	Bank	Tube	Airline
C_M	orig.	0.319**	0.206**	0.223′	0.293**	0.056	0.101**
	ER	0.186 ± 0.003	0.124 ± 0.001	0.138 ± 0.035	0.195 ± 0.009	0.053 ± 0.011	0.038 ± 0.000
$C_M^{(1)}$	orig.	0.406**	0.436**	0.289′	0.537**	0.013″	0.100**
	ER	0.244 ± 0.010	0.196 ± 0.015	0.135 ± 0.066	0.227 ± 0.038	0.053 ± 0.013	0.064 ± 0.001
$C_M^{(2)}$	orig.	0.327**	0.273**	0.198	0.349**	0.043*	0.150**
	ER	0.191 ± 0.004	0.147 ± 0.002	0.138 ± 0.040	0.203 ± 0.011	0.053 ± 0.020	0.041 ± 0.000
$C_M^{(3)}$	orig.	0.288**	0.192**	–	0.227**	0.314**	0.086**
	ER	0.165 ± 0.004	0.120 ± 0.001	–	0.186 ± 0.010	0.051 ± 0.043	0.037 ± 0.000

"Tailor Shop": Kapferer tailor-shop network ($n = 39$, $m = 4$) [41]. "Management": Krackhardt office cognitive social structure ($n = 21$, $m = 21$) [45]. "Families": Padgett Florentine families social network ($n = 16$, $m = 2$) [8]. "Bank": Roethlisberger and Dickson bank wiring-room social network ($n = 14$, $m = 6$) [65]. "Tube": The London Underground (i.e., "The Tube") transportation network ($n = 314$, $m = 14$) [66]. "Airline": Network of flights between cities, in which each layer corresponds to a single airline ($n = 3108$, $m = 530$) [57]. The rows labeled "orig." give the clustering coefficients for the original networks, and the rows labeled "ER" give the expected value and the standard deviation of the clustering coefficient in an ER random network with exactly as many edges in each layer as in the original network. For the original values, we perform a two-tailed Z-test to examine whether the observed clustering coefficients could have been produced by the ER networks. We designate the p-values as follows: $^*p < 0.05$, $^{**}p < 0.01$ for Bonferroni-corrected tests with 24 hypothesis; $'p < 0.05$, $''p < 0.01$ for uncorrected tests. We do not use any symbols for values that are not significant. We symmetrize directed networks by considering two nodes to be adjacent if there is at least one edge between them. The social networks in this table are node-aligned multiplex graphs, but the transport networks are not node-aligned. We report values that are means over different numbers of realizations: 1.5×10^5 for Tailor Shop, 1.5×10^3 for Airline, 1.5×10^4 for Management, 1.5×10^5 for Families, 1.5×10^4 for Tube, and 1.5×10^5 for Bank. Figure from [19]

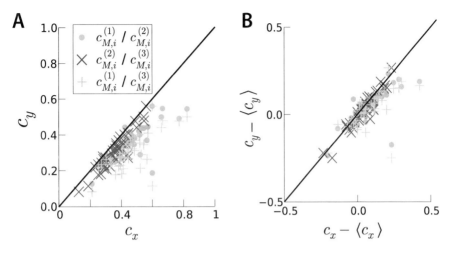

Fig. 3.3 Comparison of different local clustering coefficients in the Kapferer tailor-shop network. Each point corresponds to a node. (**a**) The raw values of the clustering coefficient. (**b**) The value of the clustering coefficients minus the expected value of the clustering coefficient for the corresponding node from a mean over 1000 realizations of a configuration model with the same degree sequence in each layer as in the original network. In a realization of the multiplex configuration model, each intra-layer network is an independent realization of the monoplex configuration model. Figure from [19]

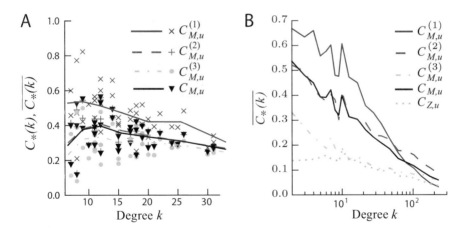

Fig. 3.4 Local clustering coefficients versus unweighted degree of the aggregated network for (**a**) the Kapferer tailor-shop network and (**b**) the airline network. The curves give the mean values of the clustering coefficients for a degree range (i.e., we bin similar degrees). Note that the horizontal axis in panel (**b**) is on a logarithmic scale. Figure from [19]

$C_{M,u}$ and the weighted clustering coefficient $C_{Z,u}$. Using a global normalization (see the discussion in Sect. 3.1.3) reduces the clustering coefficient for the small airports much more than it does for the large airports. That, in turn, introduces a bias.

The airline network is organized differently from the London Tube network. When comparing these networks, note that each layer in the former encompasses flights from a single airline. For the airline network (see Fig. 3.4b), we observe that the two-layer local clustering coefficient is larger than the single-layer one for hubs (i.e., high-degree nodes), but it is smaller for small airports (i.e., low-degree nodes). However, the global clustering coefficient counts the total number of 3-cycles and connected triplets and it thus gives more weight to high-degree nodes than to low-degree nodes, and we thus find that the global clustering coefficients for the airline network satisfy $C_M^{(2)} > C_M^{(1)} > C_M^{(3)}$. The intra-airline clustering coefficients have small values, presumably because it is not in the interest of an airline to introduce new flights between two airports that can already be reached by two flights via the same airline through some major airport. The two-layer cycles correspond to cases in which an airline has a connection from an airport to two other airports and a second airline has a direct connection between those two airports. Completing a three-layer cycle requires using three distinct airlines, and this type of congregation of airlines to the same area is not frequent in the data. Three-layer cycles are more likely than single-layer cycles only for a few of the largest airports.

The examples from empirical data show that different notions of multiplex transitivity are important in different situations. For example, the balance between intra-layer versus inter-layer clustering is different in social networks versus transportation networks (and even in different types of networks within each category, as illustrated explicitly for transportation networks), reflecting the fact that multilayer transitivity can arise from different mechanisms. Such differences are rooted in the new degrees of freedom that arise from inter-layer connections and are invisible to calculations of clustering coefficients on single-layer networks obtained via aggregation. In other words, transitivity is inherently a multilayer phenomenon: all of these diverse flavors of transitivity reduce to the same description when one throws away the multilayer information. Generalizing clustering coefficients for multiplex networks makes it possible to explore such phenomena and to gain deeper insights into different types of transitivity in networks. The existence of multiple types of transitivity also has important implications for multiplex network motifs and multiplex community structure. In particular, we can conclude that the definition of any clustering notion for multiplex networks needs to be able to handle such features.

3.3 Subgraph Centrality

In this section, we scale up the topological scale at which we consider the system and we will look at cycles and walks of all lengths. In single-layer networks, subgraph centrality is a well-established metric to measure the connectedness of a node at all scales as a generalization of the clustering coefficient that looks at a local scale. Having established the parallelism between walks in single-layer networks and supra-walks in multiplex networks, the definition of subgraph centrality and the communicability can be generalized in a direct and standardized way.

3.3.1 Subgraph Centrality, Communicability, and Estrada Index in Single-Layer Networks

The f-centrality of a node u in a single-layer network given a function f is defined as $f(\mathbf{A})_{uu}$ and the f-communicability between two distinct nodes u and v as $f(\mathbf{A})_{uv}$ [29]. Estrada and Rodríguez-Velázquez [29] defined the subgraph centrality of a node u in a single-layer network in a combinatorial way as the infinite weighted sum of closed walks of different lengths in the network starting and ending at vertex i, where the weights are the factorial of the length of each walk, i.e.

$$G_i = \sum_l \frac{\mu_i(l)}{l!} \tag{3.22}$$

where the number of cycles of length l attached to i is $\mu_i(l) = (\mathbf{A}^l)_{ii}$.

It is easy to recognize that the subgraph centrality has the following functional form:

$$G_i = (\exp(\mathbf{A}))_{ii}. \tag{3.23}$$

It coincides with the f-centrality, the function f being the exponential. Thus, the f-communicability is defined in the same way:

$$G_{ij} = (\exp(\mathbf{A}))_{ij}. \tag{3.24}$$

The matrix $\mathbf{G}$ is the communicability matrix, having on its diagonal the subgraph centrality of each node, while its off-diagonal elements encode the communicability between pairs of nodes. The Estrada index of a network is defined as

$$SC = \sum_i G_i = Tr \exp(\mathbf{A}) = Tr(\mathbf{G}) \tag{3.25}$$

3.3.2   Supra-Walks and Subgraph Centrality for Multiplex Networks

The f-centrality of node-layer pair i can be defined as a convex combination of $f(\mathcal{A}\widehat{\mathcal{C}})_{ii}$ and $f(\widehat{\mathcal{C}}\mathcal{A})_{ii}$

$$k_i = c_1 f(\mathcal{A}\widehat{\mathcal{C}})_{ii} + c_2 f(\widehat{\mathcal{C}}\mathcal{A})_{ii}, c_1, c_2 > 0. \tag{3.26}$$

Accordingly, the f-communicability between two distinct node-layer pairs i and j can be defined as a convex combination of $f(\mathcal{A}\widehat{\mathcal{C}})_{ij}$ and $f(\widehat{\mathcal{C}}\mathcal{A})_{ij}$

$$k_{ij} = c_1 f(\mathcal{A}\widehat{\mathcal{C}})_{ij} + c_2 f(\widehat{\mathcal{C}}\mathcal{A})_{ij}, c_1, c_2 > 0 \tag{3.27}$$

The supra-communicability matrix is defined as $\mathcal{K} = c_1 f(\mathcal{A}\widehat{\mathcal{C}}) + c_2 f(\widehat{\mathcal{C}}\mathcal{A})$, $c_1, c_2 > 0$. Since $f(A)^T = f(A^T)$, by using the fact that both $\mathcal{A}$ and $\widehat{\mathcal{C}}$ are symmetric, we can simplify the previous definition:

$$\mathcal{K} = c_1 f(\mathcal{A}\widehat{\mathcal{C}}) + c_2 f(\mathcal{A}\widehat{\mathcal{C}})^T. \tag{3.28}$$

As in the case of the clustering coefficient, the f-centrality of a node u results defined as the mean of the f-centrality of the node-layer pairs representing it:

$$\tilde{k}_u = \frac{1}{K_u} \sum_{i \in l(u)} k_i, \tag{3.29}$$

and the f-communicability between two distinct nodes u and v is defined as the mean of the f-communicability between each node-layer pairs representing them

$$\tilde{k}_{uv} = \frac{1}{K_u} \sum_{i \in l(u) i \in l(v)} c_1 f(\mathcal{A}\widehat{\mathcal{C}})_{ij} + \frac{1}{K_v} \sum_{i \in l(u) i \in l(v)} c_2 f(\widehat{\mathcal{C}}\mathcal{A})_{ij}. \tag{3.30}$$

We recognize that the matrix $\tilde{K} = (\tilde{k}_{uv})$ is given by:

$$\tilde{K} = c_1 \mathcal{Q}_R(f(\mathcal{A}\widehat{\mathcal{C}})) + c_2 \mathcal{Q}_L(f(\widehat{\mathcal{C}}\mathcal{A})) =$$
$$= c_1 \mathcal{Q}_R(f(\mathcal{A}\widehat{\mathcal{C}})) + c_2 \mathcal{Q}_R(f(\mathcal{A}\widehat{\mathcal{C}}))^T, \tag{3.31}$$

where $\mathcal{Q}_R(\cdot)$ (resp. $\mathcal{Q}_L(\cdot)$) is the right (resp. left) quotient associated with the partition induced by supra-nodes. For this reason, we call $\tilde{K}$ the aggregate f-communicability matrix.

This suggest $c_1 = c_2 = \frac{1}{2}$ as a good choice. It is also straightforward to see from the previous results that the following also holds[4]:

Theorem 2 *The f-centrality of a node calculated on the supra-walk matrix and the f-centrality of the same node calculated on the sum-aggregate network are equal when changing layer has no cost. The same result applies to the f-communicability between two nodes.*

---

[4] By induction, $f(\mathcal{Q}_R(A\widehat{C})) = \mathcal{Q}_L(f(\widehat{C}A))$, that is, $\tilde{\mathbf{K}} = f(\mathbf{W})$

Chapter 4
Spectra

Important information on the topological properties of a graph can be extracted from the eigenvalues of the associated adjacency, Laplacian, or any other type of graph related matrix. Thus, like spectroscopy for condensed matter physics, graph spectra are central in the study of the structural properties of a complex network.

An $N \times N$ adjacency matrix $\mathbf{A}$ is a real symmetric matrix. As such, has N real eigenvalues $\{\lambda_i\}$, $i = \ldots, N$, which we order as $\lambda_1 \leq \lambda_2 \leq \cdots \leq \lambda_N$. The set of eigenvalues with corresponding eigenvectors is unique apart from a similarity transformation, i.e., a relabeling of the nodes in the graph that obviously does not alter the structure of the graph but merely expresses the eigenvectors in a different base. Accordingly, $\mathbf{A}$ can be written as

$$\mathbf{A} = \mathbf{X} \mathbf{\Lambda} \mathbf{X}^T \tag{4.1}$$

where the $N \times N$ orthogonal matrix $\mathbf{X}$ contains, as columns, the eigenvectors $\mathbf{x}_1, \mathbf{x}_2, \ldots, \mathbf{x}_N$ of $\mathbf{A}$ belonging to the real eigenvalues $\lambda_1 \leq \lambda_2 \leq \cdots \leq \lambda_N$ and where the matrix $\mathbf{\Lambda} = \mathrm{diag}(\lambda_i)$. The eigendecomposition (4.1) is the basic relation that equates the topology (structural) domain of a network, represented by the adjacency matrix, to the spectral domain of its graph, represented by the orthogonal matrix $\mathbf{X}$ and the diagonal matrix of the eigenvalues $\mathbf{\Lambda}$.

A core subject in network theory is the connection between structure and dynamics, especially the way in which the structure affects critical phenomena. The eigendecomposition (4.1) allows to explain this connection in terms of the spectra of the adjacency matrix thus giving the basic relation that relates the topology of a network to the critical properties of the dynamics occurring on it.

© The Author(s) 2018
E. Cozzo et al., *Multiplex Networks*, SpringerBriefs in Complexity,
https://doi.org/10.1007/978-3-319-92255-3_4

4.1   The Largest Eigenvalue of the Supra-Adjacency Matrix

Consider the adjacency matrix $\mathbf{A}$ of a graph. The Perron Frobenius Theorem for nonnegative square matrices states that λ_N is simple and nonnegative, and that its associated eigenvector is the only eigenvector of $\mathbf{A}$ with nonnegative components. The largest eigenvalue λ_N is also called the spectral radius of the graph. The supra-adjacency matrix $\bar{A}$ is real and symmetric. As such, $\bar{A}$ has N real eigenvalues $\{\bar{\lambda}_i\}$, which we order as $\bar{\lambda}_1 \leq \bar{\lambda}_2 \leq \cdots \leq \bar{\lambda}_N$. Since $\bar{A}$ is also nonnegative, we have that the largest eigenvalue $\bar{\lambda}_N$ is simple and nonnegative possessing the only eigenvector of $\bar{A}$ with nonnegative components.

The largest eigenvalue of the adjacency matrix associated with a network has emerged as a key quantity for the study of a variety of different dynamical processes [55], as well as a variety of structural properties, as the entropy density per step of the ensemble of walks in a network.

In order to study the effect of the multiplexity on the spectral radius of a multiplex network, in the following we will interpret $\bar{A}$ as a perturbed version of $\mathcal{A}$, $\mathcal{C}$ being the perturbation. This choice is reasonable whenever

$$|| \mathcal{C} || < || \mathcal{A} ||, \tag{4.2}$$

where $|| \cdot ||$ is some matrix metric.

Consider the largest eigenvalue λ of $\mathcal{A}$. Since $\mathcal{A}$ is a block diagonal matrix, the spectrum of $\mathcal{A}$, $\sigma(\mathcal{A})$, is

$$\sigma(\mathcal{A}) = \bigcup_\alpha \sigma(\mathbf{A}^\alpha), \tag{4.3}$$

$\sigma(\mathbf{A}^\alpha)$ being the spectrum of the layer-adjacency matrix $\mathbf{A}^\alpha$. So, the largest eigenvalue λ of $\mathcal{A}$ is

$$\lambda = \max_\alpha \lambda_\alpha \tag{4.4}$$

with λ_α being the largest eigenvalue of $\mathbf{A}^\alpha$. We will look for the largest eigenvalue $\bar{\lambda}_N \equiv \bar{\lambda}$ of $\bar{A}$ as

$$\bar{\lambda} = \lambda + \Delta\lambda, \tag{4.5}$$

where $\Delta\lambda$ is the perturbation to λ due to the coupling $\mathcal{C}$. For this reason, the layer δ for which $\lambda_\delta = \lambda$ is named the *dominant layer*. Consider a node-aligned multiplex network. Let $\mathbf{1}_\alpha$ be a vector of size m with all entries equal to 0 except for the δth entry. If $\boldsymbol{\phi}_\delta$ is the eigenvector of $\mathbf{A}^\delta$ associated with λ_δ, we have that

$$\phi = \boldsymbol{\phi}_\delta \otimes \mathbf{1}_\alpha \tag{4.6}$$

is the eigenvector associated with λ. Observe that $\boldsymbol{\phi}_\delta$ has dimension n, while $\mathbf{1}_\alpha$ has dimension m, where n is the number of nodes, yielding to a product of dimension

$N = n \times m$. In the case in which the multiplex is not node-aligned, we must construct the vector ϕ with zeros on all positions, except on the position of the leading eigenvector of the dominant layer.

$\Delta\lambda$ can be approximated as

$$\Delta\lambda \approx \frac{\phi^T C\phi}{\phi^T \phi} + \frac{1}{\lambda}\frac{\phi^T C^2\phi}{\phi^T \phi}. \tag{4.7}$$

Because of the structure of ϕ and C, the first term on the r.h.s. is zero, while only the diagonal blocks of C^2 take part in the product $\phi^T C^2 \phi$. The diagonal blocks of C^2 are diagonals and

$$(C^2)_{ii} = \sum_{i'} C_{ii'}C_{i'i} = c_i. \tag{4.8}$$

Thus, we have that the perturbation is

$$\Delta\lambda \approx \frac{z}{\lambda}, \tag{4.9}$$

where z the weighted mean of the coupling degree with the weight given by the squares of the entries of the leading eigenvector of $\mathcal{A}$:

$$z = \sum_i c_i \frac{(\phi)_i^2}{\phi^T \phi}, \tag{4.10}$$

and it is called the *effective multiplexity*. It results that $z = 0$ in a single-layer network and $z = m - 1$ in a node-aligned multiplex network. Summing up, we have that the largest eigenvalue of the supra-adjacency matrix is equal to the largest eigenvalue of the adjacency matrix of the dominant layer at a first order approximation. As a consequence, for example, the critical point for an epidemic outbreak in a multiplex network is settled by that of the dominant layer at first order [18].

At second order, the deviation of $\bar{\lambda}$ from λ depends on the effective multiplexity and goes to zero with λ. See Figs. 4.1 and 4.2. Moreover, the approximation given in Eq. (4.9) can fail when the largest eigenvalue is near degenerated. We have two cases in which this can happen:

- the dominant layer is near degenerated,
- there is one layer (or more) with the largest eigenvalue near that of the dominant layer.

The accuracy of the approximation is related to the formula

$$\Delta\lambda \approx \phi^T C\phi + \sum_i \frac{(\phi_i^T C\phi)}{\lambda - \lambda_i}, \tag{4.11}$$

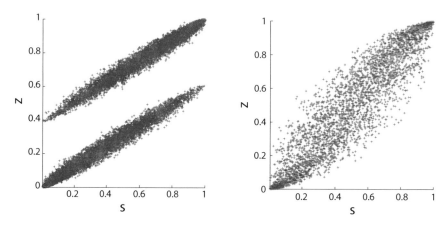

Fig. 4.1 Effective multiplexity z as a function of the fraction of nodes coupled s for a two layers multiplex with 800 nodes with a power law distribution with $\gamma = 2.3$ in each layer. For each value of s, 40 different realizations of the coupling are shown while the intra-layer structure is fixed. In the panel on the left the z shows a two band structure, while in the panel on the right, it is continuous. The difference is due to the structure of the eigenvector

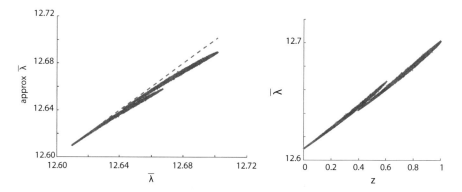

Fig. 4.2 Same setting of figure. On the right: calculated $\bar{\lambda}$. We can see two branches corresponding to the two branches of the previous figure. Left: approximated $\bar{\lambda}$ vs calculated

where λ_i and ϕ_i are the nondominant eigenvalues and the associated eigenvectors. In the first case it is evident that the second term on the r.h.s. will diverge, while in the latter, because of the structure of $\mathcal{C}$, ϕ, and ϕ_i, it is zero. In that case, we say that the multiplex network is near degenerated and we call the layers with the largest eigenvalues *codominant layers*.

When the multiplex network is near degenerated, the ϕ used in the approximation of Eq. (4.9) has a different structure. Consider that we have m_d codominant layers δ_i, $i = 1, \ldots, m_d$. If $\boldsymbol{\phi}_{\delta_i}$ is the eigenvector of $\mathbf{A}^{\delta_i}$ associated with λ_{δ_i}, we have that

$$\phi = \sum_{i=1}^{m_d} \boldsymbol{\phi}_{\delta_i} \otimes \mathbf{1}_{\delta_i}. \tag{4.12}$$

Note that the same comment on Eq. (4.6) also applies here. The term linear in C in the approximation of Eq. (4.9) is no more zero. We have

$$z_c = \frac{\phi^T C \phi}{\phi^T \phi} = \frac{1}{\phi^T \phi} \sum_{l,m:l \neq m} \phi_{\delta_l}^T \mathbf{C}_{(l,m)} \phi_{\delta_m}, \tag{4.13}$$

with $\mathbf{C}_{(l,m)}$ being the off-diagonal block (l, m) of C, and we name z_c the *correlated multiplexity*. We can decompose z_c in the contribution of each single node-layer pair

$$z_{ci} = \frac{1}{\phi^T \phi} \sum_{m:m \neq l} \sum_j \phi_{\delta_l i} C_{ij} \phi_{\delta_m j}. \tag{4.14}$$

and we call z_{ci} the *correlated multiplexity degree* of node-layer i. By definition, coupled node-layer pairs have the same correlated multiplexity degree. So, if we have m_d codominant layers in the multiplex, we get

$$\Delta \lambda \approx z_c + \frac{z}{\lambda} = m_d \sum_{i \in \delta} z_{ci} + \frac{\sum_{i \in \delta} z_i}{\lambda}. \tag{4.15}$$

4.1.1   Statistics of Walks

Given a network with adjacency matrix $\mathbf{A}$, the number of walks of length l is given by

$$N_{ij}(l) = (\mathbf{A}^l)_{ij} = \sum_r x_{ri} x_{rj} \lambda_r^l \tag{4.16}$$

where x_{ri} indicates the ith entry of the normalized eigenvector $\mathbf{x}_r$ belonging to the eigenvalue λ_r. Define the entropy $H_{ij}(l)$ of the ensemble of paths $\{\pi_{ij}(l)\}$ of length l between nodes i and j as

$$H_{ij}(l) = \ln N_{ij}(l). \tag{4.17}$$

For large walks $l \longrightarrow \infty$, it results

$$H_{ij}(l) = \ln \lambda_N + \ln(x_{Ni} x_{Nj}) \tag{4.18}$$

The leading term is independent of the positions of the endpoints. So, for large l, the entropy production rate is

$$h - \lim_{l \to \infty} \frac{H_{ij}(l)}{l} = \ln \lambda_N. \tag{4.19}$$

That is, h only depends on the largest eigenvalue of the adjacency matrix. Now, consider walks on multiplex networks that treat in the same way inter- and intra-layer steps, thus we have the supra-adjacency matrix as the multiplex walk matrix. From the perturbative approximation above, we have that the entropy production rate on a multiplex network is:

$$\bar{h} = \ln \bar{\lambda}_N \sim \ln \left(\lambda + \frac{z}{\lambda} \right) \tag{4.20}$$

That is, large walks on a multiplex network are dominated by walks on the dominant layer plus a term due to the entropy production needed to reach the dominant layer from nondominant ones.

4.2   Dimensionality Reduction and Spectral Properties

In this section, following [69], we relate the adjacency and Laplacian eigenvalues of a multiplex network to the two quotient networks we have defined in Chap. 2. The main theoretical result exploited is that the eigenvalues of a quotient interlace the eigenvalues of its parent network. Let $m < n$ and consider two sets of real numbers

$$\mu_1 \leq \cdots \leq \mu_m \quad \text{and} \quad \lambda_1 \leq \ldots \lambda_n.$$

We say that the first set *interlaces* the second if

$$\lambda_i \leq \mu_i \leq \lambda_{i+(n-m)}, \quad \text{for} \quad i = 1, \ldots, m. \tag{4.21}$$

The key spectral result is that the adjacency eigenvalues of a quotient network interlace the adjacency eigenvalues of the parent network. The same result applies for Laplacian eigenvalues, if the Laplacian matrix of the quotient is defined appropriately, i.e., as defined in Chap. 2.

4.2.1   Interlacing Eigenvalues

All the interlacing results we refer to are a consequence of the theorem below, which in turn follows from the Courant-Fisher max-min theorem

Theorem ([37, Thm. 2.1(i)]) *Let $\mathbf{A}$ be a symmetric matrix of order n, and let $\mathbf{U}$ be an $n \times m$ matrix such that $\mathbf{U}^T \mathbf{U} = \mathbf{I}$. Then the eigenvalues of $\mathbf{U}^T \mathbf{A} \mathbf{U}$ interlace those of $\mathbf{A}$.*

Observe that the matrix $\mathbf{U}^T \mathbf{A} \mathbf{U}$ is symmetric, and hence it has real eigenvalues. If $\mathbf{U}$ is the characteristic matrix of a subset $\alpha \subset \{1, 2 \ldots, n\}$, that is, $\mathbf{U} = (u_{ij})$ of size $n \times |\alpha|$ and nonzero entries $u_{i\alpha} = 1$ if $i \in \alpha$, then $\mathbf{U}^T \mathbf{A} \mathbf{U}$ equals the principal submatrix of $\mathbf{A}$ with respect to α. As $\mathbf{U}^T \mathbf{U}$ is the identity, we conclude from the theorem above:

Corollary ([37, Cor. 2.2]) *Let* $\mathbf{B}$ *be a principal submatrix of* $\mathbf{A}$. *Then the eigenvalues of* $\mathbf{B}$ *interlace the eigenvalues of* $\mathbf{A}$.

On the other hand, if $\mathbf{S}$ is the characteristic matrix of the partition, then $\mathbf{S}^T \mathbf{S} = \Lambda$ is a diagonal non-singular matrix, and hence $\mathbf{U} = \mathbf{S}\Lambda^{-1/2}$ satisfies the hypothesis of the theorem. We conclude that the eigenvalues of $\mathbf{U}^T \mathbf{A} \mathbf{U} = \Lambda^{-1/2} \mathbf{S}^T \mathbf{A} \mathbf{S} \Lambda^{-172}$ interlace those of $\mathbf{A}$. Using the Lemma 1, we conclude:

Corollary ([10, Cor. 2.3(i)]) *Let* $\mathbf{B}$ *be a quotient matrix of* $\mathbf{A}$ *with respect to some partition. Then the eigenvalues of* $\mathbf{B}$ *interlace the eigenvalues of* $\mathbf{A}$.

4.2.2 Equitable Partitions

Equation (2.22) defines equitable partitions. This can be expressed in matrix form as

$$\mathbf{AS} = \mathbf{S}Q(\mathbf{A}).$$

We call the matrix $Q(\mathbf{A})$ a regular quotient if it is the quotient of an equitable partition. If the quotient is regular, then the eigenvalues of $Q(\mathbf{A})$ not only interlace but are a subset of the eigenvalues of $\mathbf{A}$. In fact, there is a *lifting* relating both sets of eigenvalues, as we explain now.

If $\mathbf{v}$, $\mathbf{w}$ are column vectors of size m and n, we say that $\mathbf{S}\mathbf{v}$ represents the vector $\mathbf{v}$ lifted to $\mathbf{A}$, and $\mathbf{S}^T \mathbf{w}$ the vector $\mathbf{w}$ projected to $Q(\mathbf{A})$. The vector $\mathbf{S}\mathbf{v}$ has constant coordinates on each X_i, while the vector $\mathbf{S}^T \mathbf{w}$ is created by adding the coordinates on each X_i. The vector $\mathbf{w}$ is called orthogonal to the partition if $\mathbf{S}^T \mathbf{w} = 0$, that is, the sum of the coordinates over each X_i is zero. If the quotient is regular, the spectrum of $\mathbf{A}$ decomposes into the spectrum of $\mathbf{B}$ lifted to $\mathbf{A}$ (i.e., eigenvectors constant on each X_i), and the remaining spectrum is orthogonal to the partition (i.e., eigenvectors with coordinates adding to zero on each X_i):

Theorem *Let* $\mathbf{B}$ *be the quotient matrix of* $\mathbf{A}$ *with respect to an equitable partition with characteristic matrix* $\mathbf{S}$. *Then the spectrum of* $\mathbf{B}$ *is a subset of the spectrum of* $\mathbf{A}$. *More precisely,* $(\lambda, \mathbf{v})$ *is an eigenpair of* $\mathbf{B}$ *if and only if* $(\lambda, \mathbf{S}\mathbf{v})$ *is an eigenpair of* $\mathbf{A}$.

Moreover, there is an eigenbasis of $\mathbf{A}$ *of the form* $\{\mathbf{S}\mathbf{v}_1, \ldots, \mathbf{S}\mathbf{v}_m, \mathbf{w}_1, \ldots, \mathbf{w}_{(n-m)}\}$ *such that* $\{\mathbf{v}_1, \ldots, \mathbf{v}_m\}$ *is any eigenbasis of* $\mathbf{B}$, *and* $\mathbf{S}^T \mathbf{w}_i = 0$ *for all* i.

Proof The first part follows easily from the identity $\mathbf{AS} = \mathbf{SB}$ (note that $\mathbf{S}\mathbf{v} \neq 0$ as $\mathrm{Ker}(\mathbf{S}) = 0$). For the second part, note that $\mathbf{S}$ is an isomorphism onto $\mathrm{Im}(\mathbf{S})$, as it has trivial kernel, so $\{\mathbf{S}\mathbf{v}_1, \ldots, \mathbf{S}\mathbf{v}_m\}$ is a basis of $\mathrm{Im}(\mathbf{S})$. It is easy to show that the orthogonal complement $\text{Im}(\mathbf{S})^{\perp}$ equals $\text{Ker}(\mathbf{S}^T)$, hence we can complete the linearly independent set of eigenvectors $\{\mathbf{S}\mathbf{v}_1, \ldots, \mathbf{S}\mathbf{v}_m\}$ to an eigenbasis of $\mathbb{R}^n = \text{Im}(\mathbf{S}) + \text{Im}(\mathbf{S})^{\perp}$

$\square$

4.2.3   Laplacian Eigenvalues

We want to show that the Laplacian of a quotient graph is the quotient of the Laplacian matrix, as this will allow to extend the interlacing results to the Laplacian eigenvalues. First, we need to clarify what we mean by the Laplacian of a nonsymmetric matrix.

If $\mathbf{A} = (a_{ij})$ is a real symmetric (adjacency) matrix, define the *node out-degrees* as

$$d_i^{\text{out}} = \sum_j a_{ij} \text{ (row sum)}. \tag{4.22}$$

The *out-degree Laplacian* is the matrix

$$\mathbf{L}^{\text{out}} = \mathbf{D}^{\text{out}} - \mathbf{A} \tag{4.23}$$

where $\mathbf{D}^{\text{out}}$ is the diagonal matrix of the out-degrees. We define d_i^{in}, $\mathbf{D}^{\text{in}}$, and the in-degree Laplacian $\mathbf{L}^{\text{in}}$ analogously. Note that both Laplacian matrices ignore the diagonal values of $\mathbf{A}$. If $\mathbf{A}$ is the adjacency matrix of a graph, we say that the Laplacian ignores self-loops. Consider the left and right quotients of $\mathbf{A}$ with respect to a given partition. Observe that the row sums of $Q_l(\mathbf{A})$ are

$$\bar{d}_i = \frac{1}{n_i} \sum_{k \in V_i} d_k \tag{4.24}$$

the average node degree in V_i.

Let $\bar{D}$ be the diagonal matrix of the average node degrees. Then we define the *quotient Laplacian* as the matrix

$$\mathbf{L}_Q = \bar{D} - Q_l(\mathbf{A}) \tag{4.25}$$

that is, the out-degree Laplacian of the left quotient matrix. Alternatively, we could have defined L_Q as the in-degree Laplacian of the right quotient matrix, giving a transpose matrix with the same eigenvalues. (Note that there is no obvious way of interpreting the symmetric quotient $Q_s(\mathbf{L})$ as the Laplacian of a graph.) Now we can prove that the Laplacian of the quotient is the quotient of the Laplacian, in the following sense.

Theorem *Let G be a graph with adjacency matrix* $\mathbf{A}$ *and Laplacian matrix* $\mathbf{L}$. *Then:*

$$\mathbf{L}^{\text{out}}(Q_l(\mathbf{A})) = Q_l(\mathbf{L}).$$

The analogous result holds for the right quotients and the in-degree Laplacian.

Proof By definition:

$$Q_l(\mathbf{L}) = \Lambda^{-1}\mathbf{S}^T\mathbf{L}\mathbf{S} = \Lambda^{-1}\mathbf{S}^T(\mathbf{D} - \mathbf{A})\mathbf{S} = \Lambda^{-1}\mathbf{S}^T\mathbf{D}\mathbf{S} - \Lambda^{-1}\mathbf{S}^T\mathbf{A}\mathbf{S} = \bar{\mathbf{D}} - \mathbf{A}$$

The second statement follows by transposing the equation above. □

This theorem allows us to use the interlacing results of Sect. 4.2.1 for Laplacian eigenvalues. We finish by studying equitable partitions in the context of Laplacian matrices. We demonstrate that a partition being regular for the Laplacian matrix is equivalent to the partition being almost regular for the adjacency matrix. In particular, the spectral results of Sect. 4.2.2 will hold for almost regular quotients and Laplacian eigenvalues.

Theorem *Let G be a graph with adjacency matrix* **A** *and Laplacian matrix* **L**. *Then a partition is equitable with respect to* **L** *if and only if it is almost equitable with respect to* **A**.

Proof By relabeling the nodes if necessary, we can assume the block decomposition

$$A = \begin{pmatrix} A_{11} & \cdots & A_{1m} \\ \vdots & \ddots & \vdots \\ A_{m1} & \cdots & A_{mm} \end{pmatrix}, \tag{4.26}$$

where the $n_i \times n_j$ submatrix A_{ij} represents the edges from V_i to V_j. The matrix **L** has then a similar block decomposition into submatrices L_{ij}. As $\mathbf{L} = \mathbf{D} - \mathbf{A}$ and **D** is diagonal, we have $L_{ij} = -A_{ij}$ for all $i \neq j$. In particular, the row sums of L_{ij} are constant if and only if the row sums of A_{ij} are constant, for all $i \neq j$. On the other hand, as the row sums in **L** are zero, the row sums in L_{ii} equal the sum of the row sums of the matrices L_{ij} for $j \neq i$, and the result follows. □

4.3 Network of Layers and Aggregate Network

Applying the spectral results we have already presented, we conclude that the adjacency, respectively Laplacian, eigenvalues of the network of layers interlace the adjacency, respectively Laplacian, eigenvalues of the multiplex network. Namely, if $\mu_1, \ldots, \mu_m$ are the (adjacency resp. Laplacian) eigenvalues of the network of layers, then

$$\lambda_i \leq \mu_i \leq \lambda_{i+(N-m)} \quad \text{for } i = 1, \ldots, m, \tag{4.27}$$

where $\lambda_1, \ldots, \lambda_N$ are the (adjacency resp. Laplacian) eigenvalues of the multiplex network.

The network of layers, ignoring weights and self-loops, simply represents the layer connection configuration (Fig. 2.3). The connectivity of this reduced representation, measured in terms of the eigenvalues, thus relates to the connectivity of the entire multiplex network via the interlacing results.

Next, we turn to the question of when the layer partition is equitable. This requires, in particular, that the intra-layer degrees are constant, that is, each layer must be a d^α-regular graph, a very strong condition unlikely to be satisfied in real-world multiplex networks. Instead, we call a multilayer network *regular* if the layer partition is almost equitable, that is, the inter-layer connections are independent of the chosen vertices. This is a more natural condition, and it is equivalent to require the multiplex being node-aligned.

If the multiplex network is regular, i.e., node aligned, then, in addition to the interlacing, the Laplacian eigenvalues of the network of layers are a subset of the Laplacian eigenvalues of the multiplex, and we can lift a Laplacian eigenbasis of the quotient, as described in Sect. 4.2.3. This latter result has also been derived in [73] without referring to the theory of quotient graphs.

Finally, using the spectral results, we conclude that the adjacency (respectively Laplacian) eigenvalues of the aggregate network interlace the adjacency (respectively Laplacian) eigenvalues of the multiplex. Namely, in a multiplex network with N node-layer pairs and n nodes, the (adjacency resp. Laplacian) eigenvalues of the aggregate network quotient $\mu_1, \ldots, \mu_n$ satisfy

$$\lambda_i \le \mu_i \le \lambda_{i+(N-n)} \quad \text{for } i = 1, \ldots, \tilde{n}, \tag{4.28}$$

where $\lambda_1, \ldots, \lambda_N$ are the (adjacency resp. Laplacian) eigenvalues of the multiplex network. Observe that requiring the aggregate network to be regular, or almost regular, is in this case very restrictive, as it would require that every pair of nodes connects in the same uniform way on every layer, and thus it is not likely to occur on real-world multiplex networks.

The results obtained in this section will be crucial in studying structural transitions as we will show in the next chapter.[1]

4.4   Layer Subnetworks

Evidently, the layers of a multiplex form subnetworks, and it is natural to relate the eigenvalues of each layer to the eigenvalues of the multiplex. The interlacing result applies to the adjacency eigenvalues of an induced subnetwork, such as the layers, and partial interlacing also holds for the Laplacian eigenvalues. More precisely, if a layer-graph G_α has n_α nodes and adjacency (resp. Laplacian) eigenvalues

---

[1] Although here we deal only with multiplex networks, the spectral theory of quotient graphs also applies to the more general framework of multilayer networks.

$\mu_1, \ldots, \mu_\alpha$, and $\lambda_1, \ldots, \lambda_N$ are the adjacency (resp. Laplacian) eigenvalues of the whole multiplex network, then

$$\lambda_i \leq \mu_i \leq \lambda_{i+(N-n_\alpha)} \quad \text{for } i = 1, \ldots, n_\alpha, \text{ resp.} \tag{4.29}$$

$$\mu_i \leq \lambda_{i+(N-n_\alpha)} \quad \text{for } i = 1, \ldots, n_\alpha. \tag{4.30}$$

4.5   Discussion and Some Applications

From a physical point of view, the adjacency and Laplacian spectra of a network encode information on structural properties of the system represented by the network related to different dynamical processes occurring on top of it. We now discuss some consequences and applications of the spectral results derived in the previous sections. In the following, let us write $\lambda_i(\mathbf{A})$ for the ith smallest eigenvalue of a matrix $\mathbf{A}$.

4.5.1   Adjacency Spectrum

Even if the study of dynamical processes on a multiplex network is out of the scope of this book, we use some dynamical example to discuss the dimensionality results presented in the previous section. The spectrum of the adjacency matrix is directly related to different dynamical processes that take place on the system, such as spreading processes, for which it has been shown that critical properties are related to the inverse of the largest eigenvalue of this matrix. As an example, consider a contact process on the multilayer network $\mathcal{M}$ whose linearized dynamics is described by the equation

$$p_i(t+1) = \beta \sum_j \bar{a}_{ij} p_j(t) - \mu \, p_i(t), \tag{4.31}$$

in which $p_i(t)$ is the probability of node i to be infected at time t, β is the infection rate, μ is the recovery rate, and $\bar{a}_{ij}$ are the elements of the supra-adjacency matrix $\bar{A}$. In this model, each infected node contacts its neighbors with probability 1, and tries to infect them. The contact between two instances of the same object in different layers is modeled in the same way as the contact between any two other nodes.

The critical value of the disease rate for which the infection survives is given by

$$\beta_c = \frac{\mu}{\lambda_N(\bar{A})}. \tag{4.32}$$

From the interlacing result for the layer subnetworks we have that

$$\lambda_{n_\alpha}(\mathbf{A}^\alpha) \leq \lambda_N(\bar{A}), \tag{4.33}$$

This means that the critical point for the multiplex network β_c is bounded from above by the corresponding critical points of the independent layers. This implies that the multiplex network is more efficient as far as spreading processes are concerned than the most efficient of its layers on its own. On the other hand, if λ_m is the largest adjacency eigenvalue of the network of layers, then

$$\lambda_m \leq \lambda_N(\mathcal{A}), \tag{4.34}$$

which means that the connections between layers also impose constraints to the dynamics on the multiplex network. In particular, the critical point of the spreading dynamics on the multiplex network is bounded from above by the corresponding critical point of the network of layers. Interestingly, the existence of this bound explains the existence of a *mixed phase* [27].

Consider now the same process (4.31), this time defined on the aggregate network

$$p_u(t+1) = \beta \sum_v a_{uv} p_v(t) - \mu \, p_u(t). \tag{4.35}$$

Here a_{uv} are the elements of $Q(\bar{A})$, the adjacency matrix of the aggregate network. The critical value is given by

$$\widetilde{\beta}_c = \frac{\mu}{\lambda_n(Q(\bar{A}))} \tag{4.36}$$

where n is the number of nodes in $\mathcal{M}$ (the size of the aggregate network). From the interlacing result we have that

$$\widetilde{\beta}_c \geq \beta_c.$$

Therefore the spreading process on $\mathcal{M}$ is at least as effective as the same spreading process on the aggregate network. It is important to note that Eqs. (4.31) and (4.35) describe two rather different processes, that is, two different strategies that actors can adopt in order to spread information across the multiplex network. In the former, a node can infect any other node on any layer with a probability weighted by the fraction of layers in which they are in contact, while in the latter, each supra-node chooses at each time step with uniform probability a layer in which an instance representing it is present and then contacts all its neighbors in that layer. Our results show that the latter strategy is more effective than the former, as expressed by the relation between the critical points.

4.5.2   Laplacian Spectrum

The Laplacian of a network $\mathbf{L} = (l_{ij})$ is the operator of the dynamical process described by

$$\dot{p}_{ij}(t) = -\sum_k p_{ik}(t) \, l_{ki} \tag{4.37}$$

where $p_{ij}(t)$ represents the transition probability of a particle from node i to node j at time t. The second smallest eigenvalue of the Laplacian matrix sets the time scale of the process. From the interlacing results applied to the Laplacian matrix we have that for any quotient

$$\lambda_2(\bar{\mathcal{L}}) \leq \lambda_2(Q(\bar{\mathcal{L}})). \tag{4.38}$$

That is, the relaxation time on the multiplex is at most the relaxation time on any quotient, in particular the network of layers or the aggregate network. If we interpret λ_2 of the Laplacian of a network as algebraic connectivity [11], Eq. (4.38) means that the algebraic connectivity of the multiplex network is always bounded from above by the algebraic connectivity of any of its quotients.

On the other hand, the Laplacian of the aggregated network is the operator corresponding to the dynamical process described by

$$\dot{p}_{uv}(t) = \sum_w p_{uw}(t)\, a_{wv} - d_u\, p_{uv}(t) = \sum_w p_{uw}(t)\, \tilde{l}_{wu} \tag{4.39}$$

where $p_{uv}(t)$ is the transition probability of a particle from supra-node u to supra-node v at time t, a_{uw} are the elements of the adjacency matrix of the aggregated contact network, $\tilde{L} = (\tilde{l}_{ij})$ is the Laplacian matrix of the aggregate contact network (i.e., $\tilde{L} = Q(\bar{\mathcal{L}})$), and $d_u = \sum_v a_{uv}$ is the strength or degree of a node in the aggregate network). Note that if we define the overlapping degree [6] of a node as

$$o_u = \sum_v a_{uv}$$

then we have that

$$d_u = \frac{1}{\kappa_u} o_u.$$

From the interlacing result for the Laplacian we have that

$$\lambda_2(\bar{\mathcal{L}}) \leq \lambda_2(Q(\bar{\mathcal{L}})). \tag{4.40}$$

That is, the diffusion process on the aggregate network (Eq. 4.39) is faster than the diffusion process on the entire multiplex network (Eq. 4.37). Note that in [73], in a setting in which the multiplex is node-aligned, the authors obtained by means of a perturbative analysis that $\lambda_2(\bar{\mathcal{L}}) \sim \lambda_2(Q(\bar{\mathcal{L}}))$ when the diffusion parameter between layers is large enough. In [63] this result is generalized (in a different framework, since they are interested in structural properties of interdependent networks) to all almost regular multilayer networks. In the framework of quotient networks introduced in [69] and that we have presented here those results arise in a very natural way. Besides, eigenvalue interlacing between multilayer and quotient eigenvalues holds for every possible inter-layer connection scheme. In the next chapter, we will discuss the existence and location of an abrupt transition in the structure of a multiplex network by capitalizing on the interlacing results for the Laplacian. We finally note that, in the context of synchronization, the smallest nonzero Laplacian eigenvalue λ_2 is also related to the stability of a synchronized state [2], and indeed the larger λ_2 is, the more stable is the synchronized state. Considering a multiplex network, the bound in (4.38) means that the synchronized state of a system supported on the multiplex network is at most as stable as the synchronized state on any of its quotients.

4.6   The Algebraic Connectivity

The *algebraic connectivity* of a graph G is the second-smallest eigenvalue of the Laplacian matrix of G [83]. We naturally define the algebraic connectivity of a multiplex network as the second-smallest eigenvalue of its supra-Laplacian matrix. Let us call $\bar{\mu}_2$ the second-smallest eigenvalue of the supra-Laplacian and $\tilde{\mu}_2^{(a)}$ and $\tilde{\mu}_2^{(l)}$ the second-smallest eigenvalue of the aggregate and of the network of layers Laplacian, respectively. Since we are considering node-aligned multiplex networks, we have that $\tilde{mu}_2^{(l)} \equiv m$, being m the number of layers. From the interlacing results of the previous section, we know that

$$\bar{\mu}_2 \leq \tilde{\mu}_2^{(a)} \tag{4.41}$$

$$\bar{\mu}_2 \leq m \tag{4.42}$$

We also know from the inclusion relation that m is always an eigenvalue of the supra-Laplacian, so, we can look for the condition under which $\bar{\mu}_2 = m$ holds. By combining Eqs. (4.41) and (4.42), we arrive to the conclusion that

if $m \geq \tilde{\mu}_{a2}$, then $\bar{\mu}_2 \neq m$.

On the other hand, we can approximate $\bar{\mu}_2$ as

$$\bar{\mu}_2 \sim \mu_2 + \triangle\mu_2, \tag{4.43}$$

where μ_2 is the second-smallest eigenvalue of L and

$$\triangle\mu_2 = \sum_{i<j} c_{ij}(x_i - x_j)^2, \tag{4.44}$$

where $\mathbf{x}$ is the unity norm eigenvector associated with μ_2 and x_i its ith entry. Because of the structure of C and $\mathbf{x}$, it results

$$\triangle\mu_2 = m - 1, \tag{4.45}$$

for a node-aligned multiplex network. Thus, since m is always an eigenvalue of $\bar{L}$, for that approximation to be correct, the following condition must hold

$$\mu_2 + m - 1 < m, \tag{4.46}$$

from which we can conclude that

$$\text{if } \mu_2 < 1 \text{ then } \bar{\mu}_2 \neq m.$$

In summary, we have that

$$\text{if } \tilde{\mu}_2^{(a)} < m \text{ or } \mu_2 > 1 \text{ then } \bar{\mu}_2 \neq m,$$

the converse not being true in general.

This result points to a mechanism which can trigger a structural transition of a multiplex network that is different from the one exposed in [52, 63] (see next chapter).

Chapter 5
Structural Organization and Transitions

Complex networks show nontraditional critical properties due to their extreme compactness (small-world property) together with their *complex organization* [28]. The introduction of multilayer networks in general, and multiplex networks in particular, as a more natural substrate for a plethora of phenomena, poses the central theoretical question of whether critical phenomena will behave differently on such networks with respect to traditional networks. So far theoretical studies have pointed out that such differences in the critical behaviors indeed exist [60, 74]. In [63] and in [62] it has been shown that a multiplex network can exist in different *structural phases*, the transition among them being abrupt under some conditions.

In this chapter, we present how three different topological scales can be naturally identified in multiplex networks: that of the individual layers, that of the network of layers, and that of the aggregate network. The notion of quotient graph that we have introduced in Chap. 2 gives the connection between those scales in terms of the spectral properties of the parent multiplex network and of its aggregate representation.

In the rest of this chapter, we will focus on the spectra of the supra-Laplacian in order to show how the interplay between those scales affects the whole structural organization of the multiplex network. The spectrum of the Laplacian is a natural choice to address this problem since it reveals a number of structural properties. In particular, eigengaps—gaps between two subsequent eigenvalues—are known to unveil a number of structural and dynamical properties of the network related to the presence of different topological scales on it, from communities at different topological scales to synchronization patterns [3, 72]. Thus, the emerging of an eigengap points to structural changes going on, which will result in qualitatively

© The Author(s) 2018
E. Cozzo et al., *Multiplex Networks*, SpringerBriefs in Complexity,
https://doi.org/10.1007/978-3-319-92255-3_5

Fig. 5.1 Eigenvalue of a toy
multiplex with four nodes per
layers. Continuous lines are
the eigenvalues of the
multiplex networks; dashed
lines are the eigenvalues of
the aggregate network. Figure
from [17]

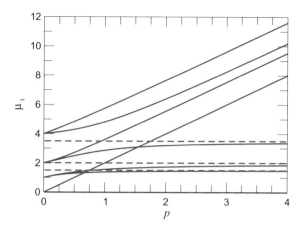

different dynamical patterns as shown in [17], which we closely follow in
Sects. 5.1, 5.2, and 5.3. We will introduce a weight parameter p for the coupling.
This parameter allows to tune the relative strength of the coupling with respect to
intra-layer connectivity.[1]

The supra-Laplacian (2.14) with the weight parameter p reads as:

$$\bar{\mathcal{L}} = \bigoplus_{\alpha} \mathbf{L}^{\alpha} + p\mathcal{L}_C, \tag{5.1}$$

and in the special case of node-aligned multiplex networks it takes the simple form:

$$\bar{\mathcal{L}} = \bigoplus_{\alpha} \left(\mathbf{L}^{(\alpha)} + p(m-1)\mathbf{I_n}\right) - p\mathbf{K_m} \otimes \mathbf{I}_n. \tag{5.2}$$

Remember that, in this special case, the spectrum of the Laplacian of the network
of layers is a subset of the spectrum of the parent supra-Laplacian. In Fig. 5.1 the
full spectrum of a toy multiplex of four nodes and two layers (then eight node-
layer pairs) is shown. The first thing to note—as observed in [35] and [73]—is
that the spectrum splits into two groups: one made up by eigenvalues that stay
bounded while increasing p, and one group of eigenvalues that diverge when
increasing p. The whole characterization of the structural changes in a multiplex
network basically depends on that splitting, i.e., on the emerging of gaps in the
spectrum.

---

[1]The weight p may have a physical meaning, like the (inverse of) commuting time in a
transportation multiplex network; however, it can be always intended as a tuning parameter.

5.1 Eigengap and Structural Transitions

The Laplacian spectrum of the network of layers is composed of just two eigenvalues: 0 with multiplicity 1, and mp with multiplicity $(m-1)$. Because of the inclusion relation between the coarse-grained and the parent spectra, mp will always be an eigenvalue of the supra-Laplacian. It results that for low enough values of p, mp will be the smallest nonzero eigenvalue of $\bar{\mathcal{L}}$. On the other hand, each eigenvalue $\bar{\mu}_i$ of $\bar{\mathcal{L}}$, with $i = 1 \ldots n$, will be bounded by the respective Laplacian eigenvalue $\tilde{\mu}_i^{(a)}$ of the aggregate network because of the interlace.

It is evident that, by increasing p, at some value $p = p^*$, it will happen that $\bar{\mu}_2 \neq mp$ and that it will approach its bound $\tilde{\mu}_2^{(a)}$. For continuity, at p^*, $\bar{\mu}_3 = mp$ must hold, since mp is always an eigenvalue of the supra-Laplacian. $p = p^*$ is the point at which the structural transition described in [52, 63] occurs, as already noted by Darabi Sahneh et al. [68]. Each eigenvalue up to $\bar{\mu}_n$ will follow the same pattern, following the line $\bar{\mu}_i = mp$ and leaving it to approach its bound $\tilde{\mu}_i^{(a)}$ when it hits the next eigenvalue $\bar{\mu}_i = mp$ (see Fig. 5.1). At the point $p = p^\diamond$ at which $\bar{\mu}_n \neq mp$, $\bar{\mu}_{n+1} = mp$ must hold and it will hold forever, since $\bar{\mu}_{n+1}$ is not bounded.

Following this reasoning, we realize that the supra-Laplacian spectrum for $p > p^\diamond$ can be divided into two groups: one of n bounded eigenvalues that will approach the aggregate Laplacian eigenvalues as p increases, and one of $N - n = n(m - 1)$ eigenvalues diverging with p. Because of that, the system can be characterized by an eigengap emerging at $p^\diamond$. Moreover, while the splitting of the eigenvalues in those two groups is always present (because of the interlacing), the crossing of the eigenvalues at p^* and at $p^\diamond$ (and between those points) only happens when the multiplex is node-aligned, since the inclusion relation only holds in that case.

In order to quantify an eigengap, we introduce the following metric:

$$g_k = \frac{\log(\bar{\mu}_{k+1}) - \log(\bar{\mu}_k)}{\log(\bar{\mu}_{k+1})} \tag{5.3}$$

and we will focus on $g_n(p)$, i.e., the gap emerging between the last bounded eigenvalue and the first unbounded at $p^\diamond$.

By construction

$$g_n(p^\diamond) = 0. \tag{5.4}$$

For $p > p^\diamond$, $\log(\bar{\mu}_{n+1})$ will diverge while $\log(\bar{\mu}_n)$ will remain bounded by $\tilde{\mu}_n^{(a)}$, so g_n will approach 1. For $p < p^\diamond$, in general both $\bar{\mu}_{n+1}$ and $\bar{\mu}_n$ will be in the continuous part of the spectrum, so g_n will be 0 in the large size limit. That is,

$$g_n = 0, \, p \leq p^\diamond$$
$$g_n \neq 0, \, p > p^\diamond. \tag{5.5}$$

Fig. 5.2 Eigengap between the last bounded and the first unbounded eigenvalue for a multiplex network of two Erdős-Rényi of 200 nodes and $< k >= 5$. Dashed line is the bound given in the text. Figure from [17]

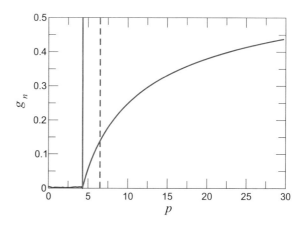

This phenomenology is confirmed by numerical experiment (see Fig. 5.2), and it describes a structural transition occurring at $p^\diamond$. In the case of a non-node-aligned multiplex network, where $p^\diamond$ is not defined since there is no crossing, $g_n(p)$ can be used to define it operationally.

An upper bound for $p^\diamond$ can be given in terms of the structural properties of the layers. If $\omega_i^{(\alpha)}$ is the strength of node u in layer α, its strength in the aggregate network is $\tilde{\omega}_i = \frac{1}{m} \sum_\alpha \omega_i^{(\alpha)}$. Next define

$$\tilde{\omega}_{ij} = \tilde{\omega}_i + \tilde{\omega}_j, \forall i \sim j \tag{5.6}$$

where $i \sim j$ indicates a link between i and j in the aggregate network. We have that [23]

$$\tilde{\mu}_n^{(a)} \leq \max_{i \sim j}\{\tilde{\omega}_{ij}\}, \tag{5.7}$$

and we can give the following bound for $p^\diamond$

$$p^\diamond \leq \frac{\max_{i \sim j}\{\tilde{\omega}_{ij}\}}{m}, \tag{5.8}$$

and

$$p^\diamond \leq \frac{\max_{i \sim j}\left\{\sum_\alpha \omega_{ij}^\alpha\right\}}{m^2} \tag{5.9}$$

The exact value of $p^\diamond$ can be derived following [68] to be

$$p^\diamond = \frac{1}{2}\lambda_n(\mathbf{Q}) \tag{5.10}$$

being, for the case of two layers, $\mathbf{Q} = \mathbf{L}^+ - \mathbf{L}^- \mathbf{L}^{+\dagger} \mathbf{L}^-$, $\mathbf{L}^\pm = \frac{1}{2}(\mathbf{L}_1 \pm \mathbf{L}_2)$, and $\mathbf{A}^\dagger$ the Moore-Penrose pseudo-inverse of $\mathbf{A}$.

5.2   The Aggregate-Equivalent Multiplex and the Structural Organization of a Multiplex Network

In order to characterize this transition, we want to compare a multiplex network $\mathcal{M}$ with the coarse-grained networks associated with it. However, a direct comparison is not possible, since those structures have different dimensionality. To overcome this problem, in [17] an auxiliary structure that has the same properties of the aggregate network and the network of layers, but also the same dimensionality of $\mathcal{M}$ is defined: the Aggregate-Equivalent Multiplex (AEM). The AEM of a parent multiplex network $\mathcal{M}$ is a multiplex network with the same number of layers of $\mathcal{M}$, each layer being identical to the aggregate network of $\mathcal{M}$. Additionally, node-layer pairs representing the same nodes are connected with a connection pattern identical to the network of layers. Formally speaking, the AEM is given by the Cartesian product between the aggregate network and the network of layers. Thus, its adjacency matrix is given by

$$\mathbf{A} = \mathbf{I}_m \otimes \tilde{\mathbf{A}} + p\mathbf{K}_m \otimes \mathbf{I}_n, \tag{5.11}$$

and its Laplacian matrix is given by

$$\mathbf{L} = \mathbf{I}_m \otimes \tilde{\mathbf{L}}_a + p\tilde{\mathbf{L}}_l \otimes \mathbf{I}_n. \tag{5.12}$$

Its Laplacian spectrum is completely determined in terms of the spectra of $\tilde{\mathbf{L}}_a$ and of the spectra of $\tilde{\mathbf{L}}_l$. In particular, we have

$$\sigma(\mathbf{L}) = \{\tilde{\mu}_a + \tilde{\mu}_l \mid \tilde{\mu}_a \in \sigma(\tilde{L}_a), \tilde{\mu}_l \in \sigma(\tilde{L}_l)\}. \tag{5.13}$$

In words, each eigenvalue of $\mathbf{L}$ is the sum of an eigenvalue of $\tilde{\mathbf{L}}_a$ and an eigenvalue of $\tilde{\mathbf{L}}_l$. We can note that since 0 is an eigenvalue of both coarse-grained Laplacians, the spectra of both $\tilde{\mathbf{L}}_a$ and $\tilde{\mathbf{L}}_l$ are included in the spectrum of $\tilde{\mathbf{L}}_a$.

To compare the parent multiplex network with its AEM, we compute the quantum relative entropies between the former and the latter. The quantum entropy (or Von-Neumann entropy) of $\mathcal{M}$ being defined as

$$S_q(\mathcal{M}) = \text{Tr}(\rho \log \rho) \tag{5.14}$$

where $\rho = \frac{\bar{\mathbf{L}}}{2E+N(m-1)p}$, with E being the number of intra-layer links in $\mathcal{M}$ [61], i.e., ρ is the supra-Laplacian normalized by the degree sum. Thus, the quantum relative entropy of the multiplex network $\mathcal{M}$ with its associated AEM is defined as

$$R_q(\mathcal{M} \parallel \text{AEM}(\mathcal{M})) = \text{Tr}\rho(\log \rho - \log \sigma), \tag{5.15}$$

with σ being the supra-Laplacian of the AEM normalized by its degree sum.

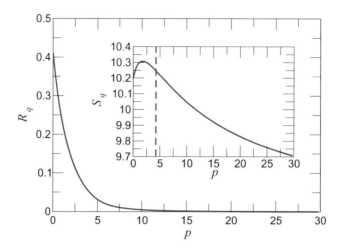

Fig. 5.3 Relative entropy ($\times 10$) (top) and Quantum Entropy (bottom) for the same system of Fig. 5.2. The vertical line indicates the exact transition point $p^\diamond$. Figure from [17]

Figure 5.3 shows the quantum relative entropy between the parent multiplex and its AEM: it goes to 0 by increasing p, meaning that the parent multiplex will be indistinguishable from the AEM. Finally, it is informative to look at the quantum entropy of $\mathcal{M}$. $S_q(\mathcal{M})$ shows a clear peak after p^* and before $p^\diamond$ (see Fig. 5.3), i.e., in the region after the transition observed in [52, 63] and before that one we have introduced here. By studying the sign of the derivative of S_q, it can be proven that the quantum entropy must have a peak before $p^\diamond$.

5.3 Dynamical Consequences and Discussions

To gain intuition on the phenomenology, it is enlightening to look at it in terms of diffusion dynamics. The large time scale is dominated by the bounded group of eigenvalues for $p \geq p^\diamond$. Those eigenvalues are close to that of the aggregate network, meaning that each layer shows practically the same behavior of the aggregate network. This is because the fast time scale is dominated by the diverging group of eigenvalues that are close to those of the aggregate network plus those of the network of layers. In summary, the network of layers determines how each node-layer pair accommodates with its replica on a fast time scale, being always "at equilibrium," while the aggregate network determines how and on what time scale the global equilibrium is attained. From that point of view, the "world" will look the same from each layer and it will look like in the aggregate network. From a random walk point of view, we can look at the average commute time $c(i, j)$, i.e., the mean time needed by a walker starting in i to hit node j for the first time and coming back. It can be expressed in terms of the eigenvalue of $\bar{L}^\dagger$, the pseudo-inverse of the supra-Laplacian. Since the eigenvalues of $\bar{L}^{\dagger}$ are the reciprocal of the eigenvalues of $\bar{L}$, the aggregate network mean commute time $\tilde{c}(i, j)$ is a good approximation of $c(i, j)$ after $p^{\diamond}$ [67]:

$$\| c(i, j) - \tilde{c}(i, j) \| \le E \frac{n(m-1)}{2p}. \tag{5.16}$$

It is interesting to note that the eigenvalues of the aggregate network do not depend on p.

Altogether, we have that before p^* the system is structurally dominated by the network of layers, whereas after $p^{\diamond}$ it is structurally dominated by the aggregate network. Between those two points the system is in an effective multiplex state, i.e., neither of the coarse-grained structures dominate. In this region, the VN-entropy—a measure of structural complexity—shows a peak. Finally, the relative entropy between the parent multiplex and its AEM varies smoothly with p, meaning that the two transitions are smooth from a global point of view.

5.4   Structural Transition Triggered by Layer Degradation

It is of interest to study the robustness of a network under degradation, that is, by random links removal (failures), deterministic links removal (attack), or the lowering of links' strength. Networks degradation models real-world processes like failures in traffic networks [84], neurodegenerative diseases [71], and many others.

Among others, the algebraic connectivity is a good graph-theoretically based measure of network robustness since it measures the extent to which it is difficult to cut the network into independent components [39]; in fact, it is a lower bound for both the edge connectivity and node connectivity of a graph, i.e., the minimal number of edges and nodes that have to be removed to disconnect the graph. Besides its structural relation to the network robustness, it is also a good measure from a dynamical point of view. For example, the time needed to synchronize a network of oscillators is related to it [2], as well as the time scale of a diffusion process. In this sense, the algebraic connectivity represents the connection between the structural and the dynamical robustness of a network.

The algebraic connectivity of a multiplex network follows two distinct regimes during the process of layer degradation too. The system experiments an abrupt structural transition as in the case of the transition experimented when varying the coupling parameter p, due to the crossing of two eigenvalues. More interestingly, unlike the structural transition observed when the coupling parameter varies, during the layer degradation it stays constant for a finite fraction of links removed as well as for a finite interval of variation of the intra-layer weights before it starts to decrease. This also differentiates the behavior of a multiplex network from that of a single-layer network.

5.5 Continuous Layers Degradation

In this section, we focus on the continuous degradation of layers connectivity. In order to model the process, we introduce a set of intra-layer weight parameters $\{t_\alpha\}$, with t_α being the weight of the links in layer α. Besides, we fix the coupling parameter p at a given value $p_0 < p^*$, such that the system is in the disconnected phase, i.e., the algebraic connectivity is $\bar{\mu}_2 = mp_0$. The supra-Laplacian now reads

$$\bar{\mathcal{L}} = \bigoplus_\alpha t_\alpha \mathbf{L}^{(\alpha)} + p_0 \mathcal{L}_C. \tag{5.17}$$

Without loss of generality, we set all the t_α's equal to 1 but one, $t_\delta = t$. In particular, we chose the layer δ as the layer with the lowest algebraic connectivity $\mu_2^{(\delta)}$ and we call it the Laplacian dominant layer (in assonance with the definition of the dominant layer given in Chap. 4).

By construction, the algebraic connectivity for $t = 1$ is $\bar{\mu}_2 = mp_0$, while the next eigenvalue can be approximated as [53]

$$\bar{\mu}_3 \sim t\mu_2^{(\delta)} + (m - 1)p_0. \tag{5.18}$$

The eigenvalue $\bar{\mu}_3$ will decrease with t and, for continuity, at a given point t^* it will hit the value mp_0. Thus, we can conclude that the algebraic connectivity follows two distinct regimes:

$$\bar{\mu}_2 = \begin{cases} mp_0, & if \ t \geq t^* \\ \sim t\mu_2^{(\delta)} + (m - 1)p_0, & if \ t < t^* \end{cases} . \tag{5.19}$$

Actually, the r.h.s. of Eq. (5.18) is an upper bound for $\bar{\mu}_3$ [75] and by equating it to mp_0 we get a lower bound for the point t^* at which the algebraic connectivity enters a distinct regime:

$$t^* > \frac{p_0}{\mu_2^{(\delta)}} \tag{5.20}$$

As we can see in Fig. 5.4, the bound (5.20) is sharp for low values of p_0, where the approximation (5.18) is good. Besides, we can appreciate that the transition only exits when $p_0 < p^*$, while for larger values the algebraic connectivity is already in the regime in which it smoothly decreases (this happens for values of p_0 larger than 0.4 in the particular setting of Fig. 5.4). We can understand this behavior as well as the mechanism that triggers the structure transition by calculating the exact value of t^* in a particular setting.

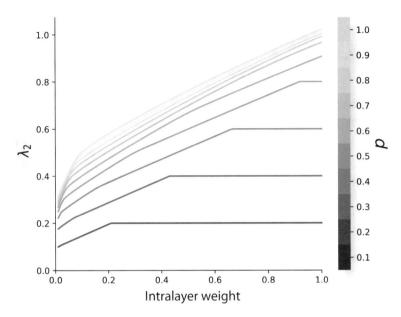

Fig. 5.4 Continuous degradation of the Laplacian dominant layer

5.5.1   *Exact Value of t* for Identical Weights*

Consider that all the intra-layer weight parameters t_α are equal, i.e., $t_\alpha = t$, $\forall \alpha$. In this case, we can give the exact value of the transition point t^* by reformulating the eigenvalue problem for the supra-Laplacian in terms of a polynomial eigenvalue problem, whose formulation will be discussed in detail in Chap. 6.

In general, a polynomial eigenvalue problem is formulated as an equation of the form

$$\mathbf{Q}(\lambda)\mathbf{x} = 0 \tag{5.21}$$

where $\mathbf{Q}$ is a polynomial matrix and the solutions are given by

$$\det(\mathbf{Q}(\lambda)) = 0 \tag{5.22}$$

The eigenvalue problem for the supra-Laplacian of a multiplex of two layers is given by

$$\bar{\mathcal{L}}\mathbf{x} = \begin{bmatrix} \mathbf{L}_a & p\mathbf{I} \\ \hline p\mathbf{I} & \mathbf{L}_b \end{bmatrix} \begin{bmatrix} \mathbf{x}_a \\ \mathbf{x}_b \end{bmatrix} = \lambda \begin{bmatrix} \mathbf{x}_a \\ \mathbf{x}_b \end{bmatrix} = \lambda \mathbf{x}. \tag{5.23}$$

From the previous expression we get the system of equations

$$\mathbf{L}_a\mathbf{x}_a - p\mathbf{I}\mathbf{x}_b = \lambda\mathbf{x}_a$$
$$\mathbf{L}_b\mathbf{x}_b - p\mathbf{I}\mathbf{x}_a = \lambda\mathbf{x}_b,$$

(5.24)

then, from the second equation we can isolate $\mathbf{x}_a$, obtaining

$$\mathbf{x}_a = -\frac{1}{p}(\mathbf{L}_b - \lambda I)\mathbf{x}_b.$$

(5.25)

Plugging the previous expression for $\mathbf{x}_a$ in the first of (5.24), we obtain

$$\left[\frac{1}{p}(\mathbf{L}_a - \lambda\mathbf{I})(\mathbf{L}_b - \lambda\mathbf{I}) - p\mathbf{I}\right]\mathbf{x}_2 = 0$$

(5.26)

in which we can recognize a quadratic eigenvalue problem

$$\mathbf{Q}(\lambda) = \mathbf{A}\lambda^2 + \mathbf{B}\lambda + \mathbf{C} = 0$$

(5.27)

with

$$\mathbf{A} = \mathbf{I}$$
$$\mathbf{B} = -(\mathbf{L}_a + \mathbf{L}_b + 2p\mathbf{I})$$
$$\mathbf{C} = \mathbf{L}_a\mathbf{L}_b + p(\mathbf{L}_a + \mathbf{L}_b).$$

(5.28)

Observe that such approach will be further explored in Chap. 6, where we will be able to easily obtain an expression for the structural transition, p^*. For more see Sect. 6.3.1 and Eq. (6.23). Complementary, if we now take into account the weights $t_a = t_b = t$ and fix the value of the coupling $p_0 = p$, Eq. (5.29) reads

$$\det(t_a\mathbf{L}_a t_b\mathbf{L}_b(t_a\mathbf{L}_a + t_b\mathbf{L}_b)^\dagger - p_0\mathbf{I}) = \det(t\mathbf{L}_a\mathbf{L}_b(\mathbf{L}_a + \mathbf{L}_b)^\dagger - p_0\mathbf{I}) = 0, \quad (5.29)$$

and it follows

$$t^* = \frac{p_0}{\lambda_2(\mathbf{L}_a\mathbf{L}_b(\mathbf{L}_a + \mathbf{L}_b)^\dagger)} = \frac{p_0}{p^*}.$$

(5.30)

In other words, we want p_0 to be the first nonzero eigenvalue of the matrix $H(t) = t\mathbf{L}_a\mathbf{L}_b(\mathbf{L}_a + \mathbf{L}_b)^\dagger$, i.e., we want the value t^* of t for which

$$p_0 = \lambda_2(H(t^*)) = t^*\lambda_2(\mathbf{L}_a\mathbf{L}_b(\mathbf{L}_a + \mathbf{L}_b)^\dagger)$$

(5.31)

from which we get Eq. (5.30).

5.5.2   General Mechanism

In general, it is always possible to write an equation of the form

$$p_0 = \lambda_2(H(\{t_\alpha^*\})) \tag{5.32}$$

that implicitly defines the point $\{t_\alpha^*\}$ of the structural transition, being this the point that solves the parameter inverse eigenvalue problem (5.32). In other words, the mechanism that triggers the structural transition is, as in the case of the transition in p, an eigenvalue crossing that results in the fact that the actual value of the coupling p is the first nonzero eigenvalue of the matrix $H(\{t\})$. For the transition in p this is obtained by directly varying it, i.e., it can be found by solving a direct eigenvalue problem, while for the transition in t this is obtained by changing the values of the intra-layer weights until p_0 is the first nonzero eigenvalue of $H(\{t\})$, i.e., it can be found by solving a parameter inverse eigenvalue problem [15]. The case of identical layers is the only one in which we can give an explicit equation for t^* as in Eq. (5.30). However, the parameter inverse eigenvalue problem can be always solved at least numerically.

In our model, the weights are constrained to be $0 < t_\alpha \leq 1$. This causes that there exists a value of the coupling parameter p_0 above which it is impossible to observe the transition in t. In fact, if we consider the Eq. (5.30) for a value of $p_0 > p^*$ we have

$$t^* = \frac{p_0}{p^*} > 1. \tag{5.33}$$

In general, given a range of variation for the intra-layer weights, it always exists a value p_c of the coupling parameter above which it is impossible to observe the transition in t. p_c can be calculated by solving

$$p_0 = \lambda_2(H(\{\bar{t}\})) \tag{5.34}$$

being $\bar{t}_\alpha$ the extremal values of t_α in its range of variation. For the setting of Fig. 5.4, in which we have two layers, one of which with fixed intra-layer weight equal to 1, we have the dependency of p_c on the extreme value of the intra-layer weight as depicted in Fig. 5.5.

5.6   Links Failure and Attacks

In this section we focus on discrete layer degradation, that is, the discrete removal of links in the multiplex network. As in the previous section, we fix one layer in which the links will be removed, picking the one with the lower algebraic connectivity,

Fig. 5.5 Dependency of p_c on the extreme value of the intra-layer weight for the same setting of Fig. 5.4

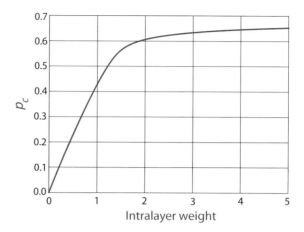

which is the Laplacian dominant layer. If we take the layer δ as isolated, when removing an edge e, its algebraic connectivity can be approximated by

$$\mu_2^{(\delta - e)} \sim \mu_2^{(\delta)} - r^2 \tag{5.35}$$

where $\mu_2^{(\delta - e)}$ is the algebraic connectivity of the layer δ with the link e removed, and $r = \left(x_{2i}^{(\delta)} - x_{2j}^{(\delta)} \right)$, where x_i is the ith entry of the eigenvector $\mathbf{x}_2^{(\delta)}$ associated with $\mu_2^{(\delta)}$.

As before, for a given $p_0 < p^*$ the algebraic connectivity of the system is $\bar{\mu}_2 = mp$, while the next eigenvalue when a link e is removed can be approximated as

$$\bar{\mu}_3 \sim \mu_2^{(\delta - e)} + (m - 1)p_0. \tag{5.36}$$

In general, when a set E of links is removed we have $r = \sum_{ij} \left(x_{2i}^{(\delta)} - x_{2j}^{(\delta)} \right)$, where the sum is over all the links in E [53]. Since Eq. (5.36) is an upper bound, as before, we obtain a lower bound for the critical value of $\mu_2^{(\delta - E)}$ from which the removal of an edge will cause a drop in the algebraic connectivity, i.e.,

$$\mu_2^{(\delta - E)*} > mp_0. \tag{5.37}$$

In this case, we have two different scenarios, the random removal of edges, which we call failures, and the targeted removal of edges, which we call attack. Let us analyze the latter scenario first. Looking at Eq. (5.36), we have two evident strategies based on the entries of the eigenvector $x_2^{(\delta)}$. We can rank the links according to the value of r associated with them and remove them in ascending or descending order. The critical fraction of edges that has to be removed in order to cause a drop in the algebraic connectivity is obviously larger in the second case, as can be observed in Fig. 5.6.

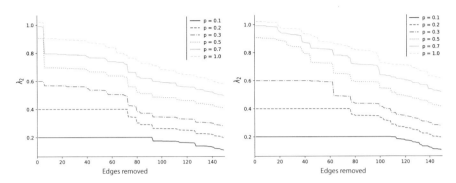

Fig. 5.6 Discrete layer degradation with descending (left panel) and ascending (right panel) ordering of the links according to p

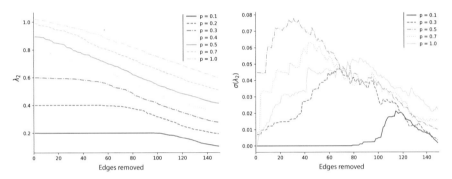

Fig. 5.7 Discrete layer degradation with random removals of links (left panel). Standard deviation (right panel)

In the case of failures, links are removed at random from the Laplacian-dominant layer. We show this result in Fig. 5.7. Moreover, in Fig. 5.7 we can also appreciate the variation of the standard deviation of the algebraic connectivity. When p_0 is low enough in order to have a structural transition, as expected, the standard deviation is constantly zero till the critical fraction of edges removed is reached.

If we look to the Fiedler eigenvector, i.e., the eigenvector associated with the algebraic connectivity, we can understand the mechanism that triggers the transition.

Without loss of generality, consider the case of a two-layer multiplex network. In the disconnected phase, namely, when $p_0 = p$, the Fiedler vector has the form $\bar{x}_2 = (1 \dots 1 \mid -1, \dots, -1)^T$, i.e., all the nodes in the same layer have the same entry of the Fiedler vector.

The structural impact on the algebraic connectivity of removing a link in a given layer can be approximated as [53]

$$\Delta \bar{\mu}_2 = \sum_{ij} (\bar{x}_{2i} - \bar{x}_{2j}) = 0, \tag{5.38}$$

which is true till $p_0 < p^*$, but the removing of a link has the side effect of lowering p^*. Thus, there will be a point in the link removal process at which p_0 is no longer lower than the actual p^*, the Fiedler vector will have a different structure and thus $\Delta\bar{\mu} \neq 0$ causing a drop in the algebraic connectivity. Note that these considerations are valid also for the continuous degradation model.

5.7   The Shannon Entropy of the Fiedler Vector

Motivated by the final considerations of the last section, we also study the Shannon entropy of the Fiedler vector, defined as

$$S = -\sum_i \bar{x}_{2i}^2 \log \bar{x}_{2i}^2, \tag{5.39}$$

considering that the Fiedler vector has a unitary norm. Moreover, as can be observed in Figs. 5.8 and 5.9, the entropy of the Fiedler vector starts at its maximum, stays constant, and experiments a discontinuous jump in correspondence with the transition. The entropy indicates the level of homogeneity of the entries of the Fiedler vector. Obviously, it is maximal when all the entries are the same, i.e., before the transition, while after that it reflects the internal organization of the multiplex network. The behavior is identical to the case of the transition in p [52], indicating that the Shannon entropy of the Fiedler vector is a good indicator of the structural transition.

5.7.1   Transition-Like Behavior for No Node-Aligned Multiplex Networks

The structural transition, both in p and in t, is due to an eigenvalue crossing between the always present eigenvalue mp_0 and the next eigenvalue. However, mp_0 is an eigenvalue of the supra-Laplacian $\bar{\mathcal{L}}$ only in the case of node-aligned multiplex networks, while for no node-aligned multiplex networks, mp_0 is only a bound [69]. This means that for no node-aligned multiplex networks a true transition doesn't exist. Interesting enough, if we perform a layer degradation by links failure we observe a similar behavior of the standard deviation as shown in Fig. 5.10.

In the initial regime, the variation is not 0 but its fluctuations are more or less constant, while in the last regime the fluctuations follow the same behavior of the true transition of the node-aligned case. Besides, we can observe a phase transition-like behavior also in the case of continuous degradation, see Fig. 5.11 in which it is shown the value of the algebraic connectivity.

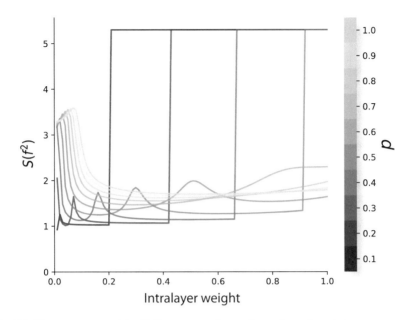

Fig. 5.8 Shannon entropy of the Fiedler vector under continuous layer degradation

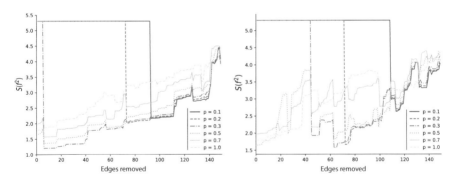

Fig. 5.9 Shannon entropy of the Fiedler vector under discrete layer degradation with descending (left panel) and ascending (right panel) ordering of the links according to p

Finally, the same happens to the Shannon entropy of the Fiedler vector (Fig. 5.12) except that now we cannot observe a jump. At variance with traditional single-layer networks, in which removing links or lowering their weights will cause in general a finite variation of the algebraic connectivity, for a multiplex network in the disconnected phase the degradation of layers will not affect the algebraic connectivity till a critical point is reached. Importantly, this is not the case for the structural transition triggered by the degradation of the coupling between layers. In this sense, multiplex networks are more resilient to damages and attacks to their layer structures than an isolated layer is if they are in the disconnected phase.

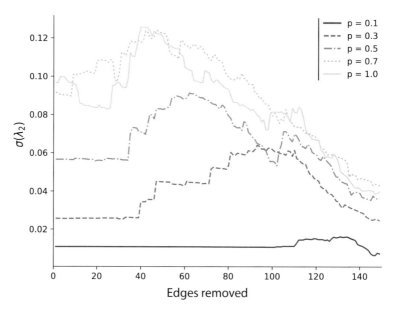

Fig. 5.10 Standard deviation of the algebraic connectivity under discrete layer degradation with random removals in a no node-aligned multiplex networks

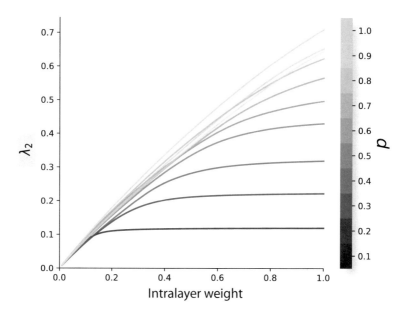

Fig. 5.11 Algebraic connectivity of a no node-aligned multiplex

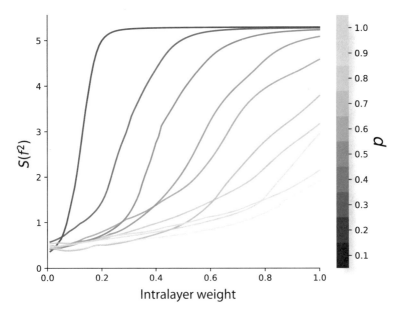

Fig. 5.12 Shannon entropy of the Fiedler vector under contentious layer degradation for a no node-aligned multiplex

The results shown here are directly applicable also to regular interdependent networks [44], since the transition is driven by an eigenvalue crossing that also applies in this case [17, 69].

Chapter 6
Polynomial Eigenvalue Formulation

In the previous chapters, we have dealt with the matricial representation for multiplex systems. Here, we focus on the matricial representation and explore the block nature of such representation. As we will show in the following, this allows us to derive interesting results based on an interpretation of the traditional eigenvalue problem. More specifically, we will reduce the dimensionality of our matrices, but increase the power of the characteristic polynomial, i.e., a polynomial eigenvalue problem. Such an approach may sound counterintuitive at first glance, but it allows us to relate the quadratic problem for a 2-Layer multiplex system with the spectra of the aggregated network and derive bounds for the spectra among many other interesting analytical insights. The main motivation for this approach can be found in [33, 46, 77]. Finally, we must also mention that this formalism is still under development. We believe it will allow us to derive new important results for multilayer network in the future. In other words, it is also a new perspective for the structural analysis of multilayer networks.

6.1 Definition of the Problem

Here we will focus our attention on 2-Layer node-aligned multiplex network composed by undirected layer-graphs. In the following sections, we will relax a little bit those assumptions, but this constraint will be useful for our first analysis of the problem. In fact, we will define the problem for a general matrix $\mathbf{M}$, in Sect. 6.1.1, and then constraint it to multiplex networks in Sect. 6.1.2.

© The Author(s) 2018
E. Cozzo et al., *Multiplex Networks*, SpringerBriefs in Complexity,
https://doi.org/10.1007/978-3-319-92255-3_6

6.1.1 Quadratic Eigenvalue Problem

In addition to the eigenvalue problem, one might also think of a higher order matricial problem. Define a matrix polynomial of degree 2, also called λ-matrix [33, 46, 77],

$$\mathbf{Q}(\lambda) = \mathbf{A}\lambda^2 + \mathbf{B}\lambda + \mathbf{C}, \tag{6.1}$$

the associated quadratic eigenvalue problem (QEP) is given by

$$\det(\mathbf{Q}(\lambda)) = 0. \tag{6.2}$$

We emphasize that a quadratic eigenvalue problem is a special case of polynomial eigenvalue problems [46], for an arbitrary matrix polynomial order in λ.

In addition to the eigenvalues, we have the right and left eigenvectors defined as

$$\mathbf{Q}(\lambda)x = 0, \tag{6.3}$$

$$y^T \mathbf{Q}(\lambda) = 0. \tag{6.4}$$

Without loss of generality, we assume in the following that the eigenvectors are unitary.

6.1.2 2-Layer Multiplex Networks

A general form of any matrix (adjacency, Laplacian or probability transition) of a multilayer network composed by two layers is given by a block matrix. Thus, the standard eigenvalue problem can be expressed as

$$\begin{bmatrix} \mathbf{M}_{11} & \mathbf{M}_{12} \\ \mathbf{M}_{21} & \mathbf{M}_{22} \end{bmatrix} \begin{bmatrix} v_1 \\ v_2 \end{bmatrix} = \lambda \begin{bmatrix} v_1 \\ v_2 \end{bmatrix}, \tag{6.5}$$

where $\mathbf{M}_{12} = \mathbf{M}_{21}^T$, since we assume undirected edges. Interpreting it as a system of equations and isolating v_1 on the first row, the dependence of the components of the eigenvector is expressed as

$$v_1 = -\mathbf{M}_{21}^{-1}(\mathbf{M}_{22} - \lambda\mathbf{I})v_2. \tag{6.6}$$

Finally, inserting it on the second row we have

$$\left[(\mathbf{M}_{11} - \lambda\mathbf{I})\mathbf{M}_{21}^{-1}(\mathbf{M}_{22} - \lambda\mathbf{I}) - \mathbf{M}_{12} \right] v_2 = 0, \tag{6.7}$$

which is a QEP, whose coefficient matrices are

$$\mathbf{A} = \mathbf{M}_{12}^{-1}, \tag{6.8}$$

$$\mathbf{B} = -\left(\mathbf{M}_{11}\mathbf{M}_{12}^{-1} + \mathbf{M}_{12}^{-1}\mathbf{M}_{22}\right), \tag{6.9}$$

$$\mathbf{C} = \mathbf{M}_{11}\mathbf{M}_{12}^{-1}\mathbf{M}_{22} - \mathbf{M}_{12} \tag{6.10}$$

In our context, exchanging $\mathbf{M}_{11}$ and $\mathbf{M}_{22}$ does not change the system, neither the solutions. However, if the polynomial considering the first layer is $\mathbf{Q}(\lambda)$, then, for the second one (relabeling the layer indices) it is $\mathbf{Q}(\lambda)^T$. In this way we found a relation between the right and left eigenvectors and these two possible configurations of our system. Formally, such observation implies $x = v_2$ and $y = v_1$. Note, however, that if the right eigenvector x is normalized, then the eigenvector of the standard eigenvalue problem, $[y \ x^T]^T$, is not normalized. In the following, we assume that x is unitary to simplify the equations. Although in this chapter we focus on static networks where only the coupling matrices are a function of a coupling parameter, all the presented formalism can be extended to more general problems. Another example of application of this formalism was shown in Sect. 5.5.1, where we analyzed the problem of layer weight degradation.

6.2 Spectral Analysis

Up to this point, we have defined our main mathematical tools, making as less constraints as possible. Now we restrict ourselves on diagonal coupling matrices and assume a linear function of the parameter p, $\mathbf{M}_{12} = p\mathbf{D}$, where $\mathbf{D}$ is a diagonal invertible matrix (such constraint will be relaxed later). Then, defining the scalar equation that describes each eigenvalue as the product of $\mathbf{Q}(\lambda)$ by its left and right eigenvectors, we have

$$y^T\mathbf{Q}(\lambda)x = a(y^T, x)\lambda^2 + b(y^T, x)\lambda + c(y^T, x) = 0, \tag{6.11}$$

where $a(y^T, x) = y^T\mathbf{A}x$, $b(y^T, x) = y^T\mathbf{B}x$, and $c(y^T, x) = y^T\mathbf{C}x$. Since this is a scalar quadratic equation, its solution is given by

$$\lambda^{\pm}(x) = \frac{-b(y^T, x) \pm \sqrt{\Delta(y^T, x)}}{2a(y^T, x)}, \tag{6.12}$$

where $\Delta(y^T, x) = b(y^T, x)^2 - 4a(y^T, x)c(y^T, x)$. Note that for each pair of right and left eigenvectors we have two possible equations, but just one of them is an eigenvalue of $\mathbf{Q}(\lambda)$. Additionally, differentiating Eq. (6.11) by p we obtain information on how the eigenvalues change as p changes. Formally, we have

$$\frac{\partial y^T\mathbf{Q}(\lambda)x}{\partial p} = y^T\frac{\partial\mathbf{Q}(\lambda)}{\partial p}x + y^T\mathbf{Q}(\lambda)\frac{dx}{dp} + \frac{dy}{dp}\mathbf{Q}(\lambda)x = 0, \tag{6.13}$$

where

$$\frac{\partial \mathbf{Q}(\lambda)}{\partial p} = 2\lambda \mathbf{D}^{-1}\frac{d\lambda}{dp} + \frac{d\lambda}{dp}\mathbf{B} + \lambda\frac{\partial \mathbf{B}}{\partial p} + \frac{\partial \mathbf{C}}{\partial p}. \tag{6.14}$$

Note that the eigenvalues and eigenvectors are also a function of p. Moreover, observe that for non-crossing points the relations $\dfrac{dy^T}{dp}\mathbf{Q}(\lambda)x = 0$ and $y^T\mathbf{Q}(\lambda)\dfrac{dx}{dp} = 0$ hold, since the derivatives are bounded for non-crossing points. However, on the crossings we have two eigenvectors associated with the same eigenvalue, which imply two solutions of the derivatives. Then, isolating the derivative of λ we have

$$\frac{d\lambda}{dp} = \frac{y^T\left(-\lambda\dfrac{\partial \mathbf{B}}{\partial p} - \dfrac{\partial \mathbf{C}}{\partial p}\right)x}{y^T\left(2\lambda \mathbf{D}^{-1} + \mathbf{B}\right)x}. \tag{6.15}$$

As an application, such relation can be used to drive a system through different regimes. For instance, considering the adjacency matrix, one can use this equation in order to choose an edge or set of edges to be removed (or weighted) in order to reduce the leading eigenvalue and consequently the critical point of spreading processes, such as epidemic spreading. The same also applies to the Laplacian. Obviously, the matrix under study depends on the process. Aside from that, another application would be the design of a numerical method to follow the right eigenvalues as a function of p.

6.2.1 Bounds

Aiming to find bounds to Eq. (6.12), we study the scalar polynomial defined by $x^T\mathbf{Q}(\lambda)x = 0$, where x is an eigenvector (left or right), which guarantees that the polynomial is equal to zero. In order to simplify the problem we multiply $\mathbf{Q}(\lambda)$ by $\mathbf{M}_{12}$, obtaining a monic polynomial matrix. Then we must bound the terms $b(x^T, x)$ and $\Delta(x^T, x)$, which will allow us to bound both solutions. Those terms can be bounded by the numerical range of the matrices in which they are related. The numerical range [38] is formally defined for any matrix $\mathbf{X}$ as $F(\mathbf{X}) = \{x^T\mathbf{X}x : x \in \mathbb{C} \text{ and } x^Tx = 1\}$. Additionally, $\sigma(\mathbf{X}) \subseteq F(\mathbf{X})$, where $\sigma(\mathbf{X})$ is in the set of eigenvalues of $\mathbf{X}$. Moreover, if $\mathbf{X}$ is an Hermitian matrix $x^T\mathbf{X}x$ is the Rayleigh quotient [38] of $\mathbf{X}$, which implies $\lambda_1(\mathbf{X}) \leq x^T\mathbf{X}x \leq \lambda_N(\mathbf{X})$. Finally, to bound a non-Hermitian matrix we use the relation of the spectral norm and the numerical range [38], given as $\frac{1}{2}|||\mathbf{X}|||_2 \leq r(\mathbf{X}) \leq |||\mathbf{X}|||_2$, where $r(\mathbf{X})$ is its numerical radius, defined as $r(\mathbf{X}) = \max\limits_{\|x\|_2=1}|x^*\mathbf{X}x| = \max\{|z| : z \in F(\mathbf{X})\}$.

Consider the term $b(x^T, x)$, which is bounded by

$$-|||\mathbf{B}|||_2 \leq b(x^T, x) \leq |||\mathbf{B}|||_2, \tag{6.16}$$

however, in many cases $\mathbf{B}$ is an Hermitian matrix, allowing us to improve this bound to

$$\lambda_{\min}(\mathbf{B}) \le b(x^T, x) \le \lambda_{\max}(\mathbf{B}) \tag{6.17}$$

More precisely, observe that $\mathbf{B}$ is often related to the aggregated network, thus enlightening the connection between both scales.

Next, we evaluate $\Delta(x^T, x) = (x^T \mathbf{B} x)^2 - 4x^T \mathbf{C} x$. Firstly, we can analyze the term $(x^T \mathbf{B} x)^2$, by observing that: (a) $\min\{\mu_i\} \le x^T \mathbf{B} x \le \max\{\mu_i\}$, (b) $\min\{\mu_i^2\} \le x \mathbf{B}^2 x \le \max\{\mu_i^2\}$, and (c) $\min\{|\mu_i|\}^2 \le (x^T \mathbf{B} x)^2 \le \max\{|\mu_i|\}^2$, since $\min\{\mu_i^2\} = \min\{|\mu_i|\}^2$, hence, from (b) and (c), bounding $(x^T \mathbf{B} x)^2$ is equivalent to bounding $x \mathbf{B}^2 x$. Secondly, we can factorize $\Delta(x^T, x) = x^T(\mathbf{B}^2 - 4\mathbf{C})x$ and defining the matrix $\Delta = \mathbf{B}^2 - 4\mathbf{C}$, we can focus on the problem $x^T \Delta x$ instead of the initial definition of $\Delta(x^T, x)$, since both have the same bounds. Besides, since in most of the problems on networks we are dealing with symmetric matrices, we might also impose that $\Delta(x^T, x) \ge 0$ because the spectra is real. In summary we have

$$0 \le \Delta(x^T, x) \le \||\Delta\||_2. \tag{6.18}$$

The bounds can be further improved when applied to the analysis of particular matrices (adjacency, Laplacian, probability transition, ...), since their particularities also impose constraints on the solutions and can be explored to improve those results.

6.2.2   Comments on Symmetric Problems: HQEP

The matrix polynomial defined by the matrix coefficients presented in (6.8) are not symmetric in most of the cases; however, a class of problems that arise naturally is defined by $\mathbf{M}_{12} = p\mathbf{I}$. In this case the matrices $\mathbf{A}$ and $\mathbf{B}$ are Hermitian; however, $\mathbf{C}$ is not. On the other hand, we can use the Toeplitz decomposition [38] in order to analyze a simplified problem. Such decomposition states that any square matrix can be uniquely written as the sum an Hermitian ($\mathbf{X} = \mathbf{X}^*$) and a skew Hermitian matrix ($\mathbf{X} = -\mathbf{X}^*$) as $\mathbf{X} = \frac{1}{2}(\mathbf{X_1} + \mathbf{X_2}^*) + \frac{1}{2}(\mathbf{X_1} - \mathbf{X_2}^*)$. It allows us to decompose $p\mathbf{C} = \frac{1}{2}(\mathbf{M}_{11}\mathbf{M}_{22} + \mathbf{M}_{22}\mathbf{M}_{11}) + \frac{1}{2}(\mathbf{M}_{11}\mathbf{M}_{22} - \mathbf{M}_{22}\mathbf{M}_{11}) + p^2\mathbf{I}$. In this way we can rewrite our QEP into two parts, one composed by Hermitian matrices, which is called Hyperbolic Quadratic Eigenvalue Problem (HQEP) [77], and a skew Hermitian matrix, that can be interpreted as a perturbation if the layers are similar. The advantage of such an approach is that HQEP presents interesting features, for instance the left and right eigenvalues coincide. The natural consequence of the perturbation theory is that the matrix $p\mathbf{C}$ of the HQEP is perturbed by $\frac{1}{2}(\mathbf{M}_{11}\mathbf{M}_{22} - \mathbf{M}_{22}\mathbf{M}_{11})$ and such a matrix norm goes to zero as the layers are more and more similar. From the Bauer and Fike theorem [38] we can write a quality function for the approximation of the perturbed matrix $\mathbf{C}$ as

$$\left|\lambda - \hat{\lambda}\right| \leq \kappa(\mathbf{U}) \left|\!\left|\!\left| \frac{1}{2}(\mathbf{M}_{11}\mathbf{M}_{22} - \mathbf{M}_{22}\mathbf{M}_{11}) \right|\!\right|\!\right|, \tag{6.19}$$

where $\hat{\lambda}$ is the eigenvalue of $\mathbf{C} = \mathbf{C_H} + \mathbf{C_S}$, $\mathbf{C_H} = \mathbf{U}\Lambda\mathbf{U}^{-1}$ and $\kappa(\cdot)$ is the condition number with respect to the matrix norm $|\!|\!|\cdot|\!|\!|$. Considering the spectral norm $|\!|\!|\cdot|\!|\!|_2$ we have $\kappa(\mathbf{X}) = \left|\frac{\sigma_{\max}(\mathbf{X})}{\sigma_{\min}(\mathbf{X})}\right|$. If $\kappa(\mathbf{U})$ is near 1, small perturbations imply small changes on the eigenvalues. On the other hand, large values of $\kappa(\mathbf{U})$ suggest a poor approximation. Observe that such an analysis concerns only the matrix $\mathbf{C}$ and not the whole QEP; however, it can be an estimate of the quality of the approximation and show that the general solution interpolates between a HQEP to a general QEP.

6.2.3 Limits for Sparse Inter-Layer Coupling

So far we assumed a node-aligned multiplex, whose coupling matrix $\mathbf{M}_{12}$ fulfill the invertibility condition, which is a necessary condition to formally define the QEP problem. However, we can use the limit of $\mathbf{D}_{ii} = \epsilon \to 0$ in order to obtain results for the sparse coupling. Observe that Eq. (6.1) can be analyzed in two different steps, first calculating the limit of decoupled nodes and next the remaining system. The first limit is analyzed as follows. From (6.1) the absent edges are factorized as

$$p^{-1}\epsilon^{-1}\tilde{\mathbf{D}}(i)\lambda^2 - p^{-1}\epsilon^{-1}\left(\mathbf{M}_{11}\tilde{\mathbf{D}}(i) + \tilde{\mathbf{D}}(i)\mathbf{M}_{22}\right)\lambda + p^{-1}\epsilon^{-1}\mathbf{M}_{11}\tilde{\mathbf{D}}(i)\mathbf{M}_{22} - p\mathbf{D} = 0, \tag{6.20}$$

where $\tilde{\mathbf{D}} = \epsilon\mathbf{D}^{-1}$. Multiplying Eq. (6.20) by $p\epsilon$ and using the following limit

$$\lim_{\epsilon \to 0}\left[\tilde{\mathbf{D}}\right]_{jj} = \begin{cases} 1 & \text{if} \quad \left[\mathbf{D}^{-1}\right]_{jj} \in O\left(\frac{1}{\epsilon}\right), \\ 0 & \text{otherwise.} \end{cases} \tag{6.21}$$

we get

$$\tilde{\mathbf{D}}\lambda^2 - \left(\mathbf{M}_{11}\tilde{\mathbf{D}} + \tilde{\mathbf{D}}\mathbf{M}_{22}\right)\lambda + \mathbf{M}_{11}\tilde{\mathbf{D}}\mathbf{M}_{22} = 0, \tag{6.22}$$

where the term of order $O(\epsilon)$ in $\mathbf{D}$ vanishes in the limit $\epsilon \to 0$. The main trick of this manipulation is to build a matrix, whose inverse does not diverge when ϵ tends to zero. In such a matrix, all the elements associated with non-coupled nodes vanish, allowing us to extract information only about part of the spectra.

Observe that $\tilde{\mathbf{D}} = \lim_{\epsilon \to 0}\left[\epsilon\mathbf{D}^{-1}\right] = \mathbf{I}$ if both layers are decoupled and the polynomial equation can be factorized as $(\mathbf{M}_{11} - \lambda\mathbf{I})(\mathbf{M}_{22} - \lambda\mathbf{I}) = 0$, whose solutions are the union of the solution of the standard eigenvalue problem of each layer. An important observation is that the number of nodes that are not connected to the other layer is also the number of eigenvalues that do not change as a function of p.

As mentioned before, Eq. (6.22) only presents the solution of the nodes that do not have any counterpart on the other layer. In order to calculate the remaining solutions we have to redefine the original problem in terms of the Moore–Penrose pseudoinverse, denoted by $\mathbf{X}^{\dagger}$, for a matrix $\mathbf{X}$. Denoting by $\bar{\mathbf{D}} = p^{-1}\mathbf{D}^{\dagger}$ we have $\bar{\mathbf{D}}_{jj} = p^{-1}\mathbf{D}_{jj}^{-1}$ if $\mathbf{D}_{jj} \neq 0$ and $\bar{\mathbf{D}}_{jj} = 0$ otherwise. Observe that the zeros of $\bar{\mathbf{D}}_{jj}$ are ones in $\tilde{\mathbf{D}}_{jj}$.

For the sake of simplicity, in the following sections we assume that $\mathbf{M}_{12}$ is invertible; however, the strategy mentioned above can be applied if it is not the case. From the computational point of view, we can reduce the cost to calculate the whole spectra as a function of a closed range of p by separating it into two components, where a subset is constant and the remaining subset varies.

6.3 Applications

Our main formalism is now developed and we will show some applications on the Laplacian and adjacency matrices in Sects. 6.3.1 and 6.3.2, respectively. In both cases, we show bounds for the spectra and also properties of the derivatives of the eigenvalues as a function of the coupling parameter p. Additionally, in the Laplacian case we were able to discuss another way to calculate the structural transition [63], however, avoiding the mathematical difficulties found in [68]. Finally, note that those matrices are related to many dynamical processes, for instance, epidemic spreading, for the adjacency matrix and diffusion and synchronization of coupled oscillators in the case of a Laplacian.

6.3.1 Supra-Laplacian Matrix

The most general Laplacian matrix considering a diagonal coupling matrix is $M_{12} = -p\mathbf{D}$, where $\mathbf{D}$ is a diagonal matrix and $\mathbf{D}_{ii} \neq 0$ (for a comment on sparse coupling, see Sect. 6.2.3). The QEP of such a matrix is defined by $\mathbf{A} = \mathbf{D}^{-1}$, $\mathbf{B} = -\left(\mathbf{L_a}\mathbf{D}^{-1} + \mathbf{D}^{-1}\mathbf{L_b} + 2p\mathbf{I}\right)$ and $\mathbf{C} = \mathbf{L_a}\mathbf{D}^{-1}\mathbf{L_b} + p\left(\mathbf{L_a} + \mathbf{L_b}\right)$. The analysis of this QEP is not trivial, since the matrices are not symmetric; however, the comparison with the diagonal coupling can give us some insights about the behavior of more general cases. For the sake of simplicity, let us consider the simplest case, where $\mathbf{D} = \mathbf{I}$. Thus, we have a monic polynomial matrix, where $\mathbf{B}$ is the aggregated network, which is semi-positive definite. Besides, $\mathbf{C}$ is a matrix that contains the product of both layers and accounts for similarities between them.

6.3.1.1 Structural Transition

Firstly, we discuss the structural transition presented in [63] on the Laplacian matrix. However, here we calculate the exact transition points using the QEP formulation. Note that we can easily derive such transition points using our formalism. It is noteworthy that those transition points were also calculated in [68] using two different methods: eigenvalue sensitivity analysis and a Shur's complement approach. Both derivations are quite complicated, contrasting with our approach, where the solutions are given using simple arguments. Note, however, that our approach yields a different expression if compared to the method presented in [68], but both expressions lead to the same final result. We do not prove this equivalence mathematically but just verified their equivalence numerically.

To begin with, we must recall that $\lambda = 2p$ is an eigenvalue of the supra-Laplacian and the crossing points are a consequence of this eigenvalue crossing the bounded part of the supra-Laplacian spectra, producing the so-called structural transitions. In this way, from our definition of QEP, we have that

$$\det\left(\mathbf{Q}(2p)\right) = \det\left(\mathbf{L_a} + \mathbf{L_b}\right) \det\left(\mathbf{L_b}\mathbf{L_a}\left(\mathbf{L_a} + \mathbf{L_b}\right)^\dagger - p\mathbf{I}\right), \tag{6.23}$$

which have two possible solutions: (1) $\det\left(\mathbf{L_a} + \mathbf{L_b}\right) = 0$, which is always true, since the sum of two Laplacian matrices is also the Laplacian of the aggregated network and also has determinant equal to zero and (2) the solution of $\det\left(\mathbf{L_b}\mathbf{L_a}\left(\mathbf{L_a} + \mathbf{L_b}\right)^\dagger - p\mathbf{I}\right)$, which are the crossing points or eigenvalues of multiplicity larger than one. Since it is also an eigenvalue problem in terms of p, we have $p^* = \lambda_i\left(\mathbf{L_b}\mathbf{L_a}\left(\mathbf{L_a} + \mathbf{L_b}\right)^\dagger\right)$. There are N possible values of p that solve (6.23), each one representing one crossing. Observe that it only crosses the N lowest eigenvalues. The first crossing is trivial, at $p = 0$, and the second one is the so-called structural transition [63], which impacts on dynamical processes. As previously mentioned, the obtained expression is different from the previous one presented in the literature, derived in [68]; however, both exhibit the same solutions as we numerically observe.

6.3.1.2 Bounds

The QEP of the supra-Laplacian can be bounded considering the individual bounds of $\mathbf{B}$, which is semi-positive definite Hermitian matrix, leading to

$$2p \leq -b(x^T, x) \leq 2p + \lambda_{\max}(\mathbf{L_a} + \mathbf{L_b}). \tag{6.24}$$

Besides, the discriminant function is also bounded by

$$\min\left\{x^T\left((\mathbf{L_a} - \mathbf{L_b})^2 - 2p\left(\mathbf{L_a} + \mathbf{L_b}\right) + 4p^2\mathbf{I}\right)x\right\} \leq \Delta(x^T, x)$$
$$\Delta(x^T, x) \leq \max\left\{x\left((\mathbf{L_a} - \mathbf{L_b})^2\right) + 4p^2\mathbf{I}\right\}, \tag{6.25}$$

where the upper bound can be defined as a function of the spectral properties of $(\mathbf{L_a} - \mathbf{L_b})^2$. On the other hand, regarding the lower bound, it can be improved by realizing that the matrix $\Delta = (\mathbf{L_a} - \mathbf{L_b})^2 - 2p\left(\mathbf{L_a} + \mathbf{L_b}\right) + 4p^2\mathbf{I}$, defined in Sect. 6.2.1, is semi-positive definite for undirected networks, $\Delta \succeq 0$. Thus, $(\mathbf{L_a} - \mathbf{L_b})^2 - 2p\left(\mathbf{L_a} + \mathbf{L_b}\right) + 4p^2\mathbf{I} \succeq 0$, hence $(\mathbf{L_a} - \mathbf{L_b})^2 + 4p^2\mathbf{I} \succeq 2p\left(\mathbf{L_a} + \mathbf{L_b}\right)$, implying that $\lambda_i \left((\mathbf{L_a} - \mathbf{L_b})^2 + 4p^2\mathbf{I}\right) \geq \lambda_i \left(2p\left(\mathbf{L_a} + \mathbf{L_b}\right)\right)$.[1] From these properties, we can establish the lower bound as $4p^2$. Formally,

$$4p^2 \leq \Delta(x^T, x) \leq \lambda_{\max}\left((\mathbf{L_a} - \mathbf{L_b})^2\right) + 4p^2 \tag{6.26}$$

The previous bounds imply that in the asymptotic analysis formalism we have $\Delta(x^T, x) \in \Theta(p^2)$. Moreover, observe that the lower and upper bounds converge to each other as the layers are similar. Finally, combining the formerly obtained bounds we have,

$$0 \leq \lambda^-(x^T, x) \leq \tfrac{1}{2}\lambda_{\max}(\mathbf{L_a} + \mathbf{L_b}) \tag{6.27}$$

and

$$2p \leq \lambda^+(x^T, x) \leq p + \frac{\lambda_{\max}(\mathbf{L_a}+\mathbf{L_b})}{2} + \frac{\sqrt{\lambda_{\max}\left((\mathbf{L_a}-\mathbf{L_b})^2\right)+4p^2}}{2}. \tag{6.28}$$

Interestingly, these bounds can be analyzed in terms of their asymptotic behavior (approximation), where for a sufficiently large value of p they can be approximated to

$$0 \leq \lambda^-(x^T, x) \leq \frac{\lambda_{\max}(\mathbf{L_a} + \mathbf{L_b})}{2}, \tag{6.29}$$

$$2p \leq \lambda^+(x^T, x) \leq 2p + \frac{\lambda_{\max}(\mathbf{L_a} + \mathbf{L_b})}{2}. \tag{6.30}$$

Observe that from the asymptotic point of view we have $\lambda^-(x) \in \Theta(1)$ and $\lambda^+(x) \in \Theta(p)$. Additionally, we can establish a condition when such bounds can be adopted.

To round up this section, we present an example in Fig. 6.1, where we present the evaluation of the eigenvalues as a function of the coupling parameter p of a multiplex network composed by two Erdös-Renyi layers with $n = 10^3$ nodes, the first layer having average degree $\langle k \rangle = 12$, while for the second $\langle k \rangle = 16$. Furthermore, we also observe that as $p \to \infty$ it also tends to the spectra of the network of layers, as expected by the interlacing properties.

6.3.1.3 Spectral Properties as a Function of the Coupling p

In this section, we analyze the spectral behavior of the supra-Laplacian matrix as a function of the coupling parameter. In other words, in this section, we explore and contextualize the analysis performed in Sect. 6.2. As usual, multilayer matrices can be understood as a function of the coupling parameter, p, allowing us to study their structural behavior as a function of this parameter.

---

[1] In addition, let us recall that if $\mathbf{M_1} - \mathbf{M_2} \succeq 0$, for semi-positive matrices $\mathbf{M_1}$ and $\mathbf{M_2}$, $\mathbf{M_1} \succeq \mathbf{M_2}$, then $\lambda_i(\mathbf{M_1}) \geq \lambda_i(\mathbf{M_2})$, where the eigenvalues are in descending order.

Fig. 6.1 Evaluation of the
eigenvalues $\lambda(\mathbf{L})$ as a
function of the coupling
parameter p of a multiplex
network composed by two
Erdös-Renyi layers with
$n = 10^3$ nodes and the first
layer have average degree
$\langle k \rangle = 12$, while the second
with $\langle k \rangle = 16$. The
continuous lines are the upper
bounds, while the dashed
lines the lower bounds

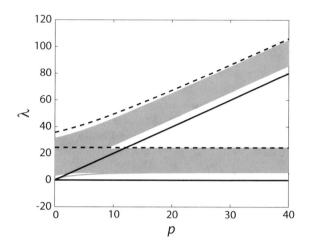

First of all, consider the simplest case, where $\mathbf{D} = \mathbf{I}$. In such a case we have
a monic polynomial matrix, where $\mathbf{B}$ is the aggregated network, which is semi-
positive definite. Besides, $\mathbf{C}$ is a matrix that contains the product of both layers and
accounts for similarities between them. Hence, (6.15) can be expressed as

$$\frac{d\lambda}{dp} = \frac{\left(2\lambda y^T x - \hat{b}(y^T, x)\right)}{\left(2(\lambda - p)y^T x - \hat{b}(y^T, x)\right)}, \qquad (6.31)$$

where $\hat{b}(y^T, x) = y^T(\mathbf{L_a} + \mathbf{L_b})x$ and $y^T x = \cos(\theta)$ is the cosine of the angle
between the left and the right eigenvectors of our QEP. Note that part of the spectra
has $\dfrac{d\lambda}{dp} \to 0$, while for the other part $\dfrac{d\lambda}{dp} \to 2$ as p increases, which can be proved
as follows. Firstly, suppose that λ is constant as a function of p, then $\dfrac{d\lambda}{dp} \to 0$
because the denominator grows as a function of p and the numerator is bounded as
supposed. Secondly, suppose that λ grows with p^r, where $r < 1$. In this case $\dfrac{d\lambda}{dp} \to$
0, by the same arguments as before, since the linear function of the denominator
dominates it. However, if $r = 1$ we have $\dfrac{d\lambda}{dp} \to 2$, since both grow linearly. Finally,
with p^r, where $r > 1$, both the numerator and the denominator are dominated by p^r,
which implies that for large p the derivative $\dfrac{d\lambda}{dp} \to 1$. This is also a contradiction,
since it was supposed to be a linear function of p. In this way, we conclude that
the derivatives of λ, for large values of p, cannot grow faster than linearly and their
growth will be one of two values, 0 or 2. Those results are in agreement with the
previously obtained bounds. Additionally, as an example, in Fig. 6.1 we can also
observe such a behavior.

In addition to the identity coupling matrix, we also evaluate the sparse coupling
case, whose analytical counterpart was presented in Sect. 6.2.3. As predicted, each

uncoupled node results in a pair of eigenvalues that does not depend on p. Due to the nature of the Laplacian matrix, where just one eigenvalue varies with p, while the other remains bounded, the set of bounded eigenvalues increases by one. For example, if we have $\tilde{n}$ uncoupled nodes, the bounded part have $n + \tilde{n}$ eigenvalues, while the "unbounded" part have $n - \tilde{n}$ eigenvalues. Note that the upper bound for the bounded part is not $\frac{1}{2}\lambda_{\max}(\mathbf{L_a} + \mathbf{L_b})$ anymore. However, the general upper bound for $\mathbf{D} = \mathbf{I}$ seems to be also an upper bound for the sparse problem, as we numerically verified in our case. The figures for these experiments are not shown since they are visually similar to Fig. 6.1.

6.3.2 Supra-Adjacency Matrix

In addition to the supra-Laplacian case, in this section we extend our analysis for the supra-adjacency matrix, whose general QEP is defined as $\mathbf{A} = \mathbf{D}^{-1}$, $\mathbf{B} = -\left(\mathbf{A}_a\mathbf{D}^{-1} + \mathbf{D}^{-1}\mathbf{A}_b\right)$ and $\mathbf{C} = \mathbf{A}_a\mathbf{D}^{-1}\mathbf{A}_b - p^2\mathbf{D}$. The analysis of such QEP is not trivial since the matrices can be asymmetric; however, the comparison with the diagonal coupling can help with the analysis of more general cases. For the sake of simplicity, let us consider the simplest case, where $\mathbf{D} = \mathbf{I}$. In such a case, we have a monic polynomial matrix, where $\mathbf{B}$ is the aggregated network. Besides, $\mathbf{C}$ is a matrix that contains the product of both layers and accounts for similarities between them. Summarizing, in this section, we follow a similar approach as previously done for the supra-Laplacian case, finding bounds for the spectra and evaluating its behavior as a function of the coupling parameter p.

6.3.2.1 Bounds

Regarding the $\mathbf{D} = \mathbf{I}$ case, we can also find bounds for the spectral distribution of the adjacency matrix. Beginning with $\mathbf{B}$, we can bound it based on its eigenvalues as

$$\lambda_{\min}(\mathbf{A_a} + \mathbf{A_b}) \leq -b(x) \leq \lambda_{\max}(\mathbf{A_a} + \mathbf{A_b}). \tag{6.32}$$

Similar to the case of the supra-Laplacian, for the discriminant we have

$$\lambda_{\min}\left((\mathbf{A_a} - \mathbf{A_b})^2\right) \leq \Delta(x^T, x) \leq \lambda_{\max}\left((\mathbf{A_a} - \mathbf{A_b})^2\right).$$

Next, combining these bounds, we can bound both solutions by

$$\frac{1}{2}\left(\lambda_{\min}(\mathbf{A_a} + \mathbf{A_b}) - \sqrt{\lambda_{\max}\left((\mathbf{A_a} - \mathbf{A_b})^2\right) + 4p^2}\right) \leq \lambda^-$$

$$\leq \frac{1}{2}\left(\lambda_{\max}(\mathbf{A_a} + \mathbf{A_b}) - \sqrt{\lambda_{\max}\left((\mathbf{A_a} - \mathbf{A_b})^2\right) + 4p^2}\right). \tag{6.33}$$

and

$$\frac{1}{2}\left(\lambda_{\min}(\mathbf{A_a} + \mathbf{A_b}) + \sqrt{\lambda_{\max}\left((\mathbf{A_a} - \mathbf{A_b})^2\right) + 4p^2}\right) \le \lambda^+$$

$$\le \frac{1}{2}\left(\lambda_{\max}(\mathbf{A_a} + \mathbf{A_b}) + \sqrt{\lambda_{\max}\left((\mathbf{A_a} - \mathbf{A_b})^2\right) + 4p^2}\right), \qquad (6.34)$$

which asymptotically converges (approximation) to

$$p \pm \frac{\lambda_{\min}(\mathbf{A_a} + \mathbf{A_b})}{2} \le \lambda^{\pm}(x) \le p \pm \frac{\lambda_{\max}(\mathbf{A_a} + \mathbf{A_b})}{2}. \qquad (6.35)$$

In other words, the spectral density of the adjacency matrix is bimodal and one part of the eigenvalues grows linearly with p, while the other part decreases at the same rate.

In Fig. 6.2 we show the previously obtained bounds for a range of values of p. Interestingly, we observe that the obtained bounds seem to present a good visual correspondence with the expected behavior of our spectral distribution. Additionally, we also note that as $p \to \infty$, it also tends to the spectra of the network of layers, as expected by the interlacing properties.

6.3.2.2 Spectral Properties as a Function of the Coupling p

Aside from the bounds, we also intend to better understand the spectral behavior as a function of p. In this section, we focus on the first derivative as a function of the coupling. From Eq. (6.15) and following a similar analysis as previously done, the first derivative is given by

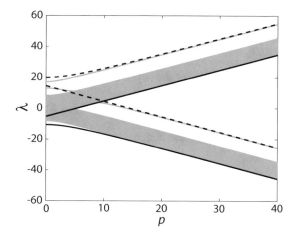

Fig. 6.2 Evaluation of the eigenvalues $\lambda(\mathbf{A})$ as a function of the coupling parameter p of a multiplex network composed by two Erdös Renyi layers with $n = 10^3$ nodes and the first layer have average degree $\langle k \rangle = 12$, while the second with $\langle k \rangle = 16$. The continuous lines are the upper bounds, while the dashed lines the lower bounds

$$\frac{d\lambda}{dp} = \frac{2py^T\mathbf{D}x}{\left(2\lambda y^T\mathbf{D}^{-1}x + b(y^T, x)\right)}, \tag{6.36}$$

where x and y^T are the right and left eigenvectors associated with λ.

Focusing on $\mathbf{D} = \mathbf{I}$ and using similar results as applied to the Laplacian case, we can suppose that λ is a constant function of p or a function of p^r with $r < 1$; however, it would give us $\frac{d\lambda}{dp} \sim p$, which is a contradiction. Next, we can suppose that it is a linear function of p, which implies $\frac{d\lambda}{dp} \to \pm 1$, depending on the sign of the linear coefficient. Finally, supposing that it is a function of p^r with $r > 1$ we obtain that $\frac{d\lambda}{dp} \to 0$, since the denominator grows faster than the numerator, which again is a contradiction. Therefore, we infer that the first derivative of λ can assume only $\frac{d\lambda}{dp} \to \pm 1$.

Next, we evaluate the sparse coupling case, whose analytical study was presented in Sect. 6.2.3. As predicted, each uncoupled node results in a pair of eigenvalues that does not depend on p. Thus, if we have $\tilde{n}$ uncoupled nodes, the central part of the spectra will have $2\tilde{n}$ eigenvalues that do not change as a function of p and $n - \tilde{n}$ that grow linearly with p, while the other $n - \tilde{n}$ eigenvalues decay with $-p$. This is illustrated in Fig. 6.3. Interestingly, also note that the obtained bound for the $\mathbf{D} = \mathbf{I}$ case seems to perform as well as on the case of Fig. 6.2, suggesting that they are also applicable in this context.

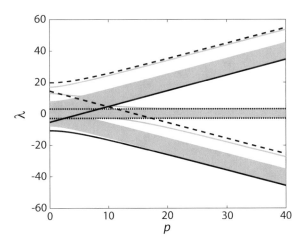

Fig. 6.3 Evaluation of the eigenvalues $\lambda(\mathbf{A})$ as a function of the coupling parameter p of a multiplex network composed by two Erdös Renyi layers with $n = 10^3$ nodes and the first layer have average degree $\langle k \rangle = 12$, while the second with $\langle k \rangle = 16$. The coupling matrix is sparse. The continuous lines is the adapted upper bound, the dashed line is the adapted lower bound, and the dotted line was obtained experimentally at the largest value of p and extended for reference

Chapter 7
Tensorial Representation

In the previous chapters, we presented and explored the matricial representation of multiplex networks. Complementary to that formalism, in this chapter, we introduce a formalism based on higher order tensors. In the case of node-aligned multiplex networks, the two representations are completely equivalent, since the supra-adjacency matrix is a particular flattening of the adjacency tensor. Tensors are elegant mathematical objects that generalize the concepts of scalars, vectors, and matrices. A tensorial representation provides a natural and concise framework for modeling and solving multidimensional problems and is widely used in different fields, from linear and multilinear algebra to physics. Here we will present the tensorial projections, and next we will analyze the spectral properties of the adjacency tensor, showing the mapping between the tensorial and the matricial representation. Finally, we show an application of this notation to obtain a concise expression for degree–degree correlation on multilayered systems in Sect. 7.4.1, allowing the characterization of different levels of correlation. The original introduction to the tensorial notation for multiplex networks is in [25]. We adopt the Einstein summation convention, in order to have more compact equations: if two indices are repeated, where one is a superscript and the other a subscript, then such operation implies a summation. Aside from that, the result is a tensor whose rank lowers by 2. For instance, $A_\beta^\alpha A_\alpha^\gamma = \sum_\alpha A_\beta^\alpha A_\alpha^\gamma$. We use Greek letters to indicate the components of a tensor. In addition, we use a tilde $(\tilde{\cdot})$ to denote the components related to the layers, with dimension m, while the components without tilde have dimension n and are related to the nodes.

© The Author(s) 2018
E. Cozzo et al., *Multiplex Networks*, SpringerBriefs in Complexity,
https://doi.org/10.1007/978-3-319-92255-3_7

7.1   Tensorial Representation

A general multilayer network is represented as the fourth-order adjacency tensor $M \in \mathbb{R}^{n \times n \times m \times m}$, which can represent several relations between nodes [25]. Formally we have

$$M_{\beta\tilde{\gamma}}^{\alpha\tilde{\delta}} = \sum_{\tilde{h},\tilde{k}=1}^{m} C_{\beta}^{\alpha}(\tilde{h}\tilde{k}) E_{\tilde{\gamma}}^{\tilde{\delta}}(\tilde{h}\tilde{k}) = \sum_{\tilde{h},\tilde{k}=1}^{m} \sum_{i,j=1}^{n} w_{ij}(\tilde{h}\tilde{k}) \mathcal{E}_{\beta\tilde{\gamma}}^{\alpha\tilde{\delta}}(ij\tilde{h}\tilde{k}), \tag{7.1}$$

where $E_{\tilde{\gamma}}^{\tilde{\delta}}(\tilde{h}\tilde{k}) \in \mathbb{R}^{m \times m}$ and $\mathcal{E}_{\beta\tilde{\gamma}}^{\alpha\tilde{\delta}}(ij\tilde{h}\tilde{k}) \in \mathbb{R}^{n \times n \times m \times m}$ indicates the tensor in its respective canonical basis.

In addition to the tensor $M \in \mathbb{R}^{n \times n \times m \times m}$, we are usually interested in a weighted tensor, in particular the case in which inter- and intra-layer edges have different weights. Such a tensor is denoted as

$$\mathcal{R}_{\beta\tilde{\delta}}^{\alpha\tilde{\gamma}}(\lambda, \eta) = M_{\beta\tilde{\sigma}}^{\alpha\tilde{\eta}} E_{\tilde{\eta}}^{\tilde{\sigma}}(\tilde{\gamma}\tilde{\delta}) \delta_{\tilde{\delta}}^{\tilde{\gamma}} + \frac{\eta}{\lambda} M_{\beta\tilde{\sigma}}^{\alpha\tilde{\eta}} E_{\tilde{\eta}}^{\tilde{\sigma}}(\tilde{\gamma}\tilde{\delta})(U_{\tilde{\delta}}^{\tilde{\gamma}} - \delta_{\tilde{\delta}}^{\tilde{\gamma}}) \tag{7.2}$$

and we call it the supra-contact tensor, whose name comes from its definition in the case of epidemic spreading [24]. Note that the intra-layer edges are weighted by η, while the inter-layer edges are weighted by λ. Furthermore, since a scalar does not change the spectral properties of our tensor—it just rescales the eigenvalues— we divide it by λ, remaining with just one parameter, the so-called coupling parameter, $\frac{\eta}{\lambda}$.

7.2   Tensorial Projections

One of the advantages of the tensorial representation is the easiness with which one can consider projections that allow to have very compact equations. In the context of multilayer networks, those projections often have a physical meaning, allowing to characterize different levels of the system. First of all, observe that we can extract one layer by projecting the tensor $M_{\beta\tilde{\gamma}}^{\alpha\tilde{\delta}}$ to the canonical tensor $E_{\tilde{\delta}}^{\tilde{\gamma}}(\tilde{r}\tilde{r})$. Formally, from Ref. [25] we have

$$M_{\beta\tilde{\gamma}}^{\alpha\tilde{\delta}} E_{\tilde{\delta}}^{\tilde{\gamma}}(\tilde{r}\tilde{r}) = C_{\beta}^{\alpha}(\tilde{r}\tilde{r}) = A_{\beta}^{\alpha}(\tilde{r}), \tag{7.3}$$

where $\tilde{r} \in \{1, 2, \ldots, m\}$ is the selected layer and $A_{\beta}^{\alpha}(\tilde{r})$ is the adjacency matrix (rank-2 tensor). Moreover, aiming at having more compact and clear equations we define the all-one tensors $u_{\alpha} \in \mathbb{R}^{n}$ and $U^{\beta\tilde{\delta}} \in \mathbb{R}^{n \times m}$. Here, we restrict our analysis to multilayer networks with diagonal couplings [44], of which mutliplex networks are a subclass. Roughly speaking, each node can have at most one counterpart on the other layers. In addition, for simplicity, we focus on unweighted and undirected connected networks in which there is a path from each node to all other nodes.

The network of layers characterizes the topology of the system in terms of the mean connectivity between layers, see Chap. 2. Formally, in tensorial notation, we have,

$$\Psi^{\tilde{\gamma}}_{\tilde{\delta}} = M^{\alpha\tilde{\gamma}}_{\beta\tilde{\delta}} U^{\beta}_{\alpha}, \tag{7.4}$$

where $\Psi^{\tilde{\gamma}}_{\tilde{\delta}} \in \mathbb{R}^{m \times m}$. Note that such a network presents self-loops, which are weighted by the number of edges in the layer. Additionally, since we assume that the layers have the same number of nodes, the edges of the network of layers have weights equal to the number of nodes n.

Another important reduction is the so-called projection [25]. Such network aggregates all the information into one layer, including self-loops that stand for the number of layers in which a node appears. Formally, we have

$$P^{\alpha}_{\beta} = M^{\alpha\tilde{\delta}}_{\beta\tilde{\gamma}} U^{\tilde{\gamma}}_{\tilde{\delta}}, \tag{7.5}$$

where $P^{\alpha}_{\beta} \in \mathbb{R}^{n \times n}$. Complementary, a version of the projection without self-edges is called the overlay network and is given as the contraction over the layers [25], i.e.

$$O^{\alpha}_{\beta} = M^{\alpha\tilde{\gamma}}_{\beta\tilde{\gamma}}. \tag{7.6}$$

Observe that the overlay network does not consider the contribution of the inter-layer connections, whereas the projection does. As we will see later, comparisons between the assortativity of those two different representations of the system reveal the key role of such inter-links.

7.3   Spectral Analysis of $\mathcal{R}(\lambda, \eta)$

A deep analysis of the spectral properties of the adjacency tensor $\mathcal{R}(\lambda, \eta)$ can inform us about the structure of multilayer networks and, consequently, give us information about dynamical processes occurring on top of such systems. First of all, the generalization of the eigenvector problem to the eigentensor is described in Sect. 7.3.1, allowing us to use some well-established linear algebra tools. Additionally, in this section, we generalize the spectral results of interlacing, obtained in [20, 69] and presented in Chap. 4, to the tensorial description adopted here. Besides, we also make use of the inverse participation ratio, IPR(Λ), as a measurement of eigenvalue localization [24, 34], a well-known tool to study the eigenvectors, which also provides important insights for epidemic processes. As a convention, we assume that the eigenvalues are ordered as $\Lambda_1 \geq \Lambda_2 \geq \ldots \geq \Lambda_{nm}$ and the individual layer eigenvalues are denoted as Λ_i^l. Note that, in contrast to Chap. 4, here we use a different eigenvalue order.

7.3.1 Eigenvalue Problem

The usual eigenvalue problem can be generalized to the case of a rank-4 tensor. In the case of the supra-contact tensor, it reads

$$\mathcal{R}^{\alpha\tilde{\gamma}}_{\beta\tilde{\delta}} f_{\alpha\tilde{\gamma}}(\Lambda) = \Lambda f_{\beta\tilde{\delta}}(\Lambda), \tag{7.7}$$

where Λ is an eigenvalue and $f_{\beta\tilde{\delta}}(\Lambda)$ is the corresponding eigentensor. In addition, we are assuming that the eigentensors form an orthonormal basis. Importantly, the supra-contact matrix, $\bar{R}$, in [18] can be understood as a flattening of the tensor $\mathcal{R}^{\alpha\tilde{\gamma}}_{\beta\tilde{\delta}}(\lambda, \eta)$. Consequently, all the results for $\bar{R}$ also apply to the tensor $\mathcal{R}$. As argued in [25], that supra-adjacency matrix corresponds to unique unfolding of the fourth-order tensor M yielding square matrices. Following this unique mapping we have the correspondence of the eigensystems.

7.3.2 Inverse Participation Ratio

In addition to the eigenvalues, we can study the behavior of the eigenvectors in terms of their localization properties, i.e., how the entries of a normalized eigenvector are distributed. One way to study this phenomenon is using the inverse participation ratio, which is also used in the context of epidemic spreading in [24, 34]. Mathematically, the inverse participation ratio is defined as

$$\mathrm{IPR}(\Lambda) \equiv \left(f_{\beta\tilde{\delta}}(\Lambda) \right)^4 U^{\beta\tilde{\delta}}. \tag{7.8}$$

In the limit of $nm \to \infty$, if the IPR(Λ) is of order $\mathcal{O}(1)$, then the eigentensor is localized and the components of $f_{\beta\tilde{\delta}}(\Lambda)$ are of order $\mathcal{O}(1)$ only for a few nodes. On the other hand, if IPR$(\Lambda) \to 0$ then this state is delocalized and the components of $f_{\beta\tilde{\delta}}(\Lambda) \sim \mathcal{O}\left(\frac{1}{\sqrt{nm}}\right)$. Additionally, another possible scenario completely different from the traditional single layer one is possible if we consider a layer-wise localization [24], i.e., localization on layers, instead of on a fraction of nodes. In such a case, the $IPR(\Lambda)$ will be of order $\mathcal{O}(1/n)$ in the localized phase, whereas it will be of order $\mathcal{O}(1/nm)$ in the delocalized phase. A deeper analysis of this scenario is presented in Sect. 7.3.5.5.

7.3.3   Interlacing Properties

Invoking the unique mapping presented on the previous subsection and considering the results of Sánchez-García et al. and Cozzo et al. [20, 69], we can use the interlacing properties to relate the spectra of the multilayer network with the spectra of the network of layers. First, we define the normalized network of layers (see Sect. 7.2) in terms of the supra contact tensor as

$$\Phi_{\tilde{\delta}}^{\tilde{\gamma}}(\lambda, \eta) = \frac{1}{n}\mathcal{R}_{\beta\tilde{\delta}}^{\alpha\tilde{\gamma}}(\lambda, \eta)U_{\alpha}^{\beta}, \tag{7.9}$$

where we are implicitly assuming a multilayer network in which the layers have the same number of nodes (the demonstration that the matrix used in [69] is an unfolding of such a tensor is shown in Sect. 7.3.4 in order to have a more fluid text). Additionally, let's denote by $\mu_1 \geq \mu_2 \geq \ldots \geq \mu_m$ the ordered eigenvalues of $\Phi_{\tilde{\delta}}^{\tilde{\gamma}}(\lambda, \eta)$. Following [69], the interlacing properties imply

$$\Lambda_{nm-m+j} \leq \mu_j \leq \Lambda_j, \tag{7.10}$$

for $j = m, \ldots, 1$. As examples, Table 7.1 shows the spectrum of three simple networks of layers that can be computed analytically: a line with two and three nodes and a triangle. Figure 7.1 shows a schematic illustration of those three multilayer networks.

Furthermore, using similar arguments we can also obtain results for the normalized projection, formally given as

$$\mathbf{P}_{\beta}^{\alpha} = \frac{1}{m}\mathcal{R}_{\beta\tilde{\delta}}^{\alpha\tilde{\gamma}}(\lambda, \eta)U_{\tilde{\gamma}}^{\tilde{\delta}}, \tag{7.11}$$

Table 7.1 Structure and spectra of the normalized network of layers $\Phi_{\tilde{\delta}}^{\tilde{\gamma}}(\lambda, \eta)$

Network	$\Phi_{\tilde{\delta}}^{\tilde{\gamma}}(\lambda, \eta)$	Eigenvalues
Line with two nodes	$\begin{bmatrix} \langle k^{l=1}\rangle & \frac{\eta}{\lambda} \\ \frac{\eta}{\lambda} & \langle k^{l=2}\rangle \end{bmatrix}$	$\langle k\rangle - \frac{\eta}{\lambda}$ $\langle k\rangle + \frac{\eta}{\lambda}$
Line with three nodes	$\begin{bmatrix} \langle k^{l=1}\rangle & \frac{\eta}{\lambda} & 0 \\ \frac{\eta}{\lambda} & \langle k^{l=2}\rangle & \frac{\eta}{\lambda} \\ 0 & \frac{\eta}{\lambda} & \langle k^{l=3}\rangle \end{bmatrix}$	$\langle k\rangle$ $\langle k\rangle - \sqrt{2}\frac{\eta}{\lambda}$ $\langle k\rangle + \sqrt{2}\frac{\eta}{\lambda}$
Multiplex	$\begin{bmatrix} \langle k^{l=1}\rangle & \frac{\eta}{\lambda} & \frac{\eta}{\lambda} \\ \frac{\eta}{\lambda} & \langle k^{l=2}\rangle & \frac{\eta}{\lambda} \\ \frac{\eta}{\lambda} & \frac{\eta}{\lambda} & \langle k^{l=3}\rangle \end{bmatrix}$	$\langle k\rangle - \frac{\eta}{\lambda}$ $\langle k\rangle - \frac{\eta}{\lambda}$ $\langle k\rangle + 2\frac{\eta}{\lambda}$

Assuming that the average degree of each layer, $\langle k^l\rangle$, is the same, i.e., $\langle k^l\rangle = \langle k\rangle, \forall l$

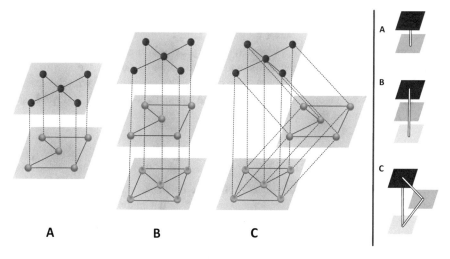

Fig. 7.1 Schematic illustration of the three multilayer networks cases considered as examples. Top panels represent the original networks which give rise to three distinct configurations for the networks of layers. See the text for more details

whose ordered eigenvalues, denoted by $\nu_1 \geq \nu_2 \geq \ldots \geq \nu_m$, also interlace those of the supra-contact tensor satisfying

$$\Lambda_{nm-n+j} \leq \nu_j \leq \Lambda_j, \tag{7.12}$$

for $j = n, \ldots, 1$. Finally, the eigenvalues of the adjacency tensor of an extracted layer also interlaces those of the parent multilayer networks, yielding

$$\Lambda_{nm-n+j} \leq \Lambda_j^l \leq \Lambda_j, \tag{7.13}$$

for $j = n, \ldots, 1$. These results show that the largest eigenvalue of the multilayer adjacency tensor is always larger than or equal to all of the eigenvalues of the individual isolated layers as well as that of the network of layers.

7.3.4 Proof of Eq. (7.9)

In this subsection we show the equivalence of the results presented in [20, 69], presented here in Chap. 4, and Eq. (7.9). Consider the matricial representation of a multilayer network, given by

$$\bar{A} = \bigoplus_{\alpha} \mathbf{A}^{\alpha} + \mathcal{C} = \begin{bmatrix} \mathbf{A}^1 & \mathbf{C}^{12} & \cdots & \mathbf{C}^{1m} \\ \mathbf{C}^{21} & \mathbf{A}^2 & \cdots & \mathbf{C}^{2m} \\ \vdots & \vdots & \ddots & \vdots \\ \mathbf{C}^{m1} & \mathbf{C}^{m2} & \cdots & \mathbf{A}^m \end{bmatrix} \tag{7.14}$$

where $\bar{A} \in \mathbb{R}^{nm \times nm}$, $\mathbf{A}^{\alpha} \in \mathbb{R}^{n \times n}$ is the adjacency matrix of the layer $\alpha \in \{1, 2, \ldots m\}$ and $\mathcal{C}$ is a coupling matrix (see Chap. 2). Since we assume multilayer networks in which the layers have the same number of nodes we have $\mathbf{C}^{\alpha\beta} = I$ if layers α and β are coupled. We assume a partition of such network, represented by $S \in \mathbb{R}^{nm \times m}$, which is the characteristic matrix of such partition, where $S_{ij} = 1$ if $i \in V_j$ and zero otherwise (V_j is the partition of the network of layers).

In order to use the results of Sánchez-García et al. and Cozzo et al. [20, 69] one has to prove that the network of layers contact matrix $\tilde{R}$ is an unfolding of our tensor $\Phi_{\tilde{\delta}}^{\tilde{\gamma}}(\lambda, \eta)$, with $\tilde{R}$ formally given by

$$\tilde{R} = \Gamma^{-1} S^T \bar{A} S, \tag{7.15}$$

where Γ is a diagonal matrix with normalizing constants (for more, see Ref. [20, 69]). In words, the product $\bar{A}S$ is a summation over the blocks of the matrix $\bar{A}$, resulting in a matrix with the degree of each node. The subsequent left product with S^T imposes another summation, whose result is a matrix composed by the sum of all elements of the blocks. Finally, the product by Γ^{-1} normalizes the result by $\frac{1}{n}$. Formally we have,

$$\bar{A}S = \begin{bmatrix} k^{11} & k^{12} & \cdots & k^{1m} \\ k^{21} & k^{22} & \cdots & k^{2m} \\ \vdots & \vdots & \ddots & \vdots \\ k^{m1} & k^{m2} & \cdots & k^{mm} \end{bmatrix} \tag{7.16}$$

where $k^{ij} \in \mathbb{R}^{n \times 1}$ is a vector with the number of edges emanating from each node on layer i to layer j and $\bar{A}S \in \mathbb{R}^{nm \times m}$. Then,

$$S^T \bar{A}S = \begin{bmatrix} \sum k^{11} & \sum k^{12} & \cdots & \sum k^{1m} \\ \sum k^{21} & \sum k^{22} & \cdots & \sum k^{2m} \\ \vdots & \vdots & \ddots & \vdots \\ \sum k^{m1} & \sum k^{m2} & \cdots & \sum k^{mm} \end{bmatrix} \tag{7.17}$$

where $\sum k^{ij} \in \mathbb{R}$ are scalars with the number of edges that connect a node on layer i to a node on layer j. Finally, the product by Γ^{-1} introduces the average degree instead of the summation, producing the same results as Eq. (7.9).

7.3.5   2-Layer Multiplex Case

In the next sections, we analyze the spectral properties of simple multilayer networks varying the configuration of the layers.

7.3.5.1 Eigenvalue Crossing

Let us analyze the spectra of a simple setup: multiplex networks composed by l identical layers. Such class of networks provides insights about the spectral behavior as a function of $\left(\frac{\eta}{\lambda}\right)$. Although they are not very realistic a priori, there are situations in which this representation is helpful: for instance, in the context of disease contagion, one might think of a multi-strain disease in which each strain propagates in a different layer allowing coinfection of the host population.

The supra contact tensor can be written as

$$\mathcal{R}^{\alpha\tilde{\gamma}}_{\beta\tilde{\delta}}(\lambda,\eta) = A^{\alpha}_{\beta}\delta^{\tilde{\gamma}}_{\tilde{\delta}} + \frac{\eta}{\lambda}\delta^{\alpha}_{\beta}K^{\tilde{\gamma}}_{\tilde{\delta}}, \tag{7.18}$$

where A^{α}_{β} is the 2-rank layer adjacency tensor, $K^{\tilde{\delta}}_{\tilde{\gamma}}$ is the adjacency tensor of the network of layers, which is a complete graph on the multiplex case, and δ^{α}_{β} is the Kronecker delta. Observe that in this case the supra-adjacency matrix is given by a Kronecker product, see Chap. 2. In this way, the eigenvalue problem can be written as

$$\mathcal{R}^{\alpha\tilde{\gamma}}_{\beta\tilde{\delta}} f_{\alpha\tilde{\gamma}} = A^{\alpha}_{\beta}\delta^{\tilde{\gamma}}_{\tilde{\delta}} f_{\alpha\tilde{\gamma}} + \frac{\eta}{\lambda}\delta^{\alpha}_{\beta}K^{\tilde{\gamma}}_{\tilde{\delta}} f_{\alpha\tilde{\gamma}}, \tag{7.19}$$

where the sum of the eigenvalues of A, Λ^{l}_{i}, and K, μ_{i}, are also eigenvalues of the adjacency tensor, hence $\mathcal{R}^{\alpha\tilde{\gamma}}_{\beta\tilde{\delta}} f_{\alpha\tilde{\gamma}} = \left(\Lambda^{l}_{i} + \frac{\eta}{\lambda}\mu_{j}\right) f_{\alpha\tilde{\gamma}}$, $i = 1, 2, \ldots n$ and $j = 1, 2, \ldots m$. Then,

$$\left(\Lambda^{l}_{i} + \frac{\eta}{\lambda}\mu_{j}\right) = \left(\Lambda^{l}_{k} + \frac{\eta}{\lambda}\mu_{s}\right). \tag{7.20}$$

The eigenvalues of the complete graph on m nodes are $\mu_{1} = m-1$, and $\mu_{i} = -1$, $\forall i > 1$, yielding

$$\frac{\eta}{\lambda} = \frac{\Lambda^{l}_{k} - \Lambda^{l}_{i}}{m}, \tag{7.21}$$

which imposes crossings on the eigenvalues of the adjacency tensor for identical layers, since $\left(\frac{\eta}{\lambda}\right)$ is a continuous parameter.

7.3.5.2 Identical Layers

Consider a multiplex network made up of two layers with the same configuration. Each layer of the multiplex is a network composed by $n = 1000$, $\langle k \rangle \approx 6$, $\Lambda^{l} = 14.34$, with degree distribution $P(k) \sim k^{-2.7}$. Aside from the intra-edge configuration, we also impose that inter-edges connect a node with its counterpart on the other layer, i.e., every node has the same intra-layer degree on all layers. Such

Fig. 7.2 Spectral properties of the tensor $\mathcal{R}(\lambda, \eta)$ as a function of the ratio $\frac{\eta}{\lambda}$ for a multiplex with two layers with the exact same degree distribution and connected to its counterpart on the other layer. On the top panel we present the inverse participation ratio (IPR(Λ)) of the three larger eigenvalues, while on bottom panel we show the leading eigenvalues. Every curve is composed by 10^3 log spaced points, in order to have enough resolution

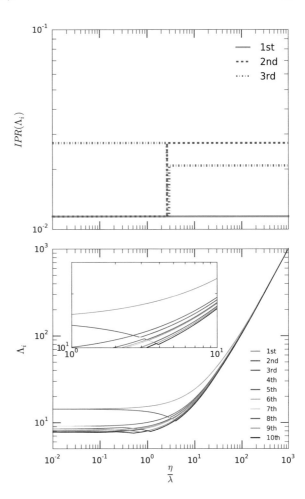

a constraint imposes a high correlation between the degree of nodes in different layers, in fact, it is the maximal correlation possible given the degree distribution.

Figure 7.2 shows the behavior of the spectra of such a multiplex network as a function of the parameter $\left(\frac{\eta}{\lambda}\right)$. In the top panel, we represent the inverse participation ratio associated with the first three eigenvalues, while on the bottom panel, we plot the first ten eigenvalues. When the ratio $\frac{\eta}{\lambda} = 0$ the eigenvalues have multiplicity two, as can be seen on the left side of the bottom panel (approximately, since the figure starts from 10^{-2}). More importantly, those eigenvalues tend to behave differently: one increases, while the other tends to decrease. This behavior leads to an eigenvalues crossing. The inset of the bottom panel zooms out the region where the crossing takes place. Note that the eigenvalues cross at the same value for which the inverse participation ratio shows an abrupt change. Indeed, the jump in the IPR(Λ) has its roots in the interchange of the eigenvectors associated with each of the eigenvalues that are crossing. Moreover, we stress that the abrupt change observed for IPR(Λ) is always present in such scenarios, but it could be either from the lower to the higher values or vice versa depending on the structure of the layers.

7.3.5.3   Similar Layers

In addition to the identical case, we have also considered a multiplex network composed by two layers with the same degree distribution (i.e., the same degree sequence), with $P(k) \sim k^{-2.7}$, but different random realizations of the configuration model. Furthermore, the inter-edges follow the same rule as before, connecting nodes with their counterparts on the other layer assuring that every node has the same intra-degree on all layers. Each layer of the multiplex network is composed by $n = 1000$ and $\langle k \rangle \approx 6$. Since each layer is a different realization of the configuration model, both present a slightly different leading eigenvalue, the first $\Lambda_1^1 = 15.21$ and the second $\Lambda_1^2 = 14.34$.

Figure 7.3 shows the spectral behavior of such a multiplex network in terms of the largest eigenvalues, on the bottom panel, and the IPR(Λ), on the top panel. Here, in addition to the global inverse participation ratio, we also present the contribution of each layer to this measure. Such analysis is meaningless on the identical case since the contribution is the same. As shown in the figure, we observe that for small values of $\frac{\eta}{\lambda}$, with respect to the first eigenvalue, the system is localized on the first layer and delocalized on the second. On the other hand, the picture changes when we focus on the second eigenvalue, as it is localized on the second layer, but delocalized on the first one. For larger values of $\frac{\eta}{\lambda}$, both layers contribute equally to IPR(Λ). Analogously to the identical case, there is a change on IPR(Λ_2), which seems to be related to the changes on Λ_2, as one can see on the bottom panel and in the inset. Note that for this case, there is no crossing, i.e., the eigenvalues avoid the crossing—also referred to as near-crossing.

7.3.5.4   Different Layer

In this section, we focus on the case of two completely different layer structures, with spaced leading eigenvalues. Consider a multiplex network made up of two scale-free networks with $\gamma \approx 2.2$ and $\gamma \approx 2.8$. Both layers have $\langle k \rangle \approx 8$ and $n = 10^3$ nodes on each layer and the leading eigenvalues are $\Lambda_1^1 = 42.64$ for the first and $\Lambda_1^2 = 21.29$ for the second.

Figure 7.4 shows the spectral properties of the tensor $\mathcal{R}(\lambda, \eta)$ as a function of the ratio $\frac{\eta}{\lambda}$. In contrast to the identical layers, Sect. 7.3.5.2, and the case of statistically equivalent layers, Sect. 7.3.5.3 (also see Figs. 7.2 and 7.3), where some eigenvalues increase while others decrease, here all the observed eigenvalues always increase.

Fig. 7.3 Spectral properties of the tensor $\mathcal{R}(\lambda, \eta)$ as a function of the ratio $\frac{\eta}{\lambda}$ for a multiplex with two layers with the same degree distribution (different random realizations of the configuration model) and connected to its counterpart on the other layer. On the top panel we present the inverse participation ratio (IPR(Λ)) of the two larger eigenvalues and the individual layer contributions, while on bottom panel we show the leading eigenvalues. Every curve is composed by 10^3 log spaced points, in order to have enough resolution

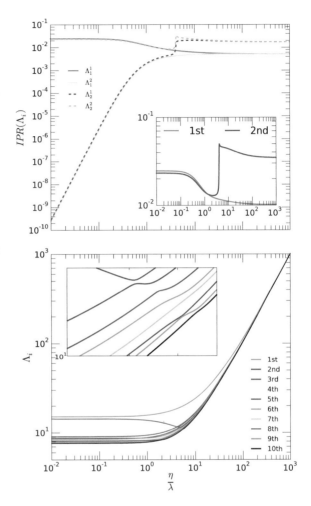

Moreover, we do not observe any crossing or near-crossing behavior. Regarding IPR(Λ), the same pattern as for the case of similar multilayer configuration is found: for small values of $\frac{\eta}{\lambda}$ and considering the first eigenvalue, the system appears localized on the first layer and delocalized on the second, while for IPR(Λ_2), it is the contrary. For larger values of $\frac{\eta}{\lambda}$, both layers contribute equally to the IPR(Λ). Furthermore, the main difference we observe for the current setup with respect to the two similar networks (see Fig. 7.3, presented in Sect. 7.3.5.3) is that now no drastic change on the inverse participation ratio is found, as expected since there is no near-crossing.

Fig. 7.4 Spectral properties of the tensor $\mathcal{R}(\lambda, \eta)$ as a function of the ratio $\frac{\eta}{\lambda}$ for a multiplex with two layers, the first with $\gamma \approx 2.2$, while the second $\gamma \approx 2.8$. Both have $\langle k \rangle \approx 8$. On the top panel we present the inverse participation ratio (IPR(Λ)) of the two larger eigenvalues and the individual layer contributions, while on bottom panel we show the leading eigenvalues. Every curve is composed by 10^3 log spaced points, in order to have enough resolution

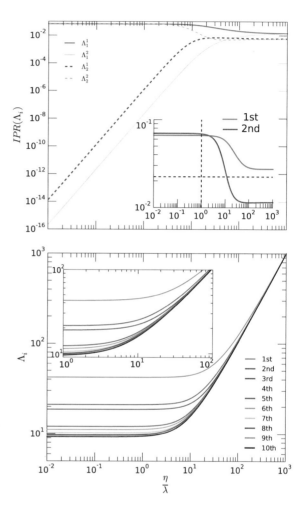

7.3.5.5   Layer-Wise Localization

One of the main results of de Arruda et al. [24] is the layer-wise localization phenomena. As mentioned previously in Sect. 7.3.2, in such a case, the $IPR(\Lambda)$ will be of order $\mathcal{O}(1/n)$ in the localized phase, whereas it will be of order $\mathcal{O}(1/nm)$ in the delocalized phase. This is because in the layer-wise localized phase the components of the eigentensor are of order $\mathcal{O}(1/\sqrt{n})$ for all the nodes in the dominant layer and of order zero for nodes in other layers. Noting that, one easily realizes that the correct finite-size scaling analysis to take in order to characterize such a transition is $m \rightarrow \infty$, i.e., the number of layers goes to infinity while the number of nodes per layer stays constant. In fact, in this limit $IPR(\Lambda)$ will vanish on one side of the transition point while remaining finite on the other side. In this way, we can observe localized states also in the case in which there is no possibility for localization in each of the layers if isolated [24].

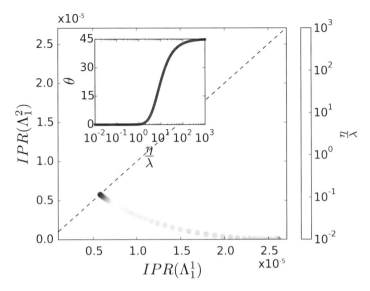

Fig. 7.5 Diagram of the contribution of each layer to the IPR(Λ) for different values of the spreading ratio $\frac{\eta}{\lambda}$. The dashed line represents the case where both layers have the same contribution, i.e., a line with slope one. On the inset we show the angle θ between the vector composed by the contributions of each layer to the IPR(Λ), $v = \left[\text{IPR}(\Lambda_1^1), \text{IPR}(\Lambda_1^2)\right]^T$, and the x-axis. The multiplex network used here is composed of two Erdös-Rényi networks, both with $n = 5 \times 10^4$, the first layer $\langle k \rangle = 16$ $((\Lambda_1^1)^{-1} \approx 0.0625)$, while the second $\langle k \rangle = 12$ $((\Lambda_1^2)^{-1} \approx 0.0833)$

Complementary, Fig. 7.5 shows the contribution of each layer to the IPR(Λ) considering different values $\frac{\eta}{\lambda}$. Calculations were performed over a multiplex network composed by two Erdös-Rényi networks, both with $n = 5 \times 10^4$, the first layer $\langle k \rangle = 16$, while the second $\langle k \rangle = 12$. Observe that for lower values of $\frac{\eta}{\lambda}$ the main contribution comes from one layer, configuring a layer-wise localized state and consequently placed on one the x-axis of Fig. 7.5. Then, increasing the ratio $\frac{\eta}{\lambda}$ we also increase the inverse participation ratio of the second layer, however decreasing the inverse participation ratio of the first layer, implying that the points tend to be on the diagonal line with slope one, where the contributions of both IPRs are the same. Such observation is also confirmed by the angle, θ, between the vector composed by the IPR contributions, $v = \left[\text{IPR}(\Lambda_1^1), \text{IPR}(\Lambda_1^2)\right]^T$, and the x-axis, where we observe it changing from 0° to 45°.

7.3.6   3-Layer Multilayer Case

Following the main ideas of the last sections, we explore the spectral properties in multilayer networks with more than two layers. Specifically, we have carried out numerical simulations for a 3-layer system. We generate multilayer networks using three scale-free networks, with $\gamma \approx 2.3$, $\gamma \approx 2.6$, and $\gamma \approx 2.9$, with $\langle k \rangle \approx 8$

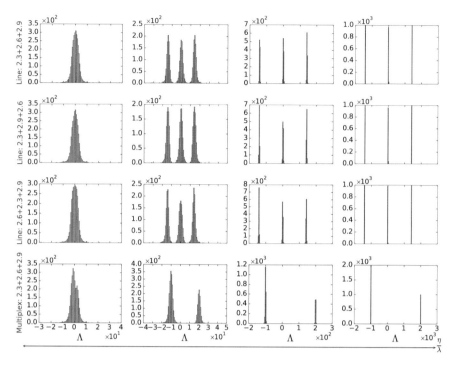

Fig. 7.6 Distribution of the eigenvalues. On the rows, from top to bottom, for the interconnected networks of Lines 2.3+2.6+2.9, 2.3+2.9+2.6, 2.6+2.3+2.9, and the multiplex. On the columns, from left to right, we varied the ratios $\frac{\eta}{\lambda} = 1$, 10, 100, and 1000, respectively. All histograms were built with 100 bins

and $n = 10^3$ nodes on each layer. Note that we have two possible topologies for the network of layers: (1) a line graph and (2) a triangle (which is a node-aligned multiplex). In its turn, the first can be arranged in three possible configurations by changing the central layer. That is, we have four possible systems. Consequently, the structure of the network of layers imposes itself more strongly on the eigenvalues of the entire interconnected structure.

7.3.6.1 Spectral Distribution

As in [24], where the authors evaluated the spectral distribution as a function of the coupling parameter, we present the same results in this section. Figure 7.6 shows the spectrum of the four configurations of networks when varying the ratio $\frac{\eta}{\lambda} = 1$, 10, 100, and 1000. Observe that we do not show the ratio $\frac{\eta}{\lambda} = 0$ since it is just the union of the individual layers' spectrum. For $\frac{\eta}{\lambda} = 1$, the four configurations are very similar, especially the line graphs. In such case, the inter-layer edges are treated in the same way as the intra-layer ones. In other words, they are ignored and the network can be interpreted as a monoplex network. As the spreading ratio increases, the spectrum tends to be clustered near the values of the eigenvalues of the network of layers. Such spectra was analytically calculated in Sect. 7.3.3 and shown in Table 7.1 on the main text.

Regarding the triangle configuration, the clustering of the spectrum as $\frac{\eta}{\lambda}$ increases is even clear. Triangles present the lowest eigenvalue with multiplicity two. On the extreme case of $\frac{\eta}{\lambda} \gg 1$, see Fig. 7.6, we have 2/3 of the values near the left extreme value while 1/3 is near the leading eigenvalue. On the other hand, for the line configurations, the frequencies of the eigenvalues distribution is related to the position of the central layer. However, in the limiting cases, such differences are reduced. This pattern is naturally related to the increase of the spreading ratio: When $\frac{\eta}{\lambda}$ increases so does the role of the inter-layer edges relative to the intra-layer connections. Consequently, the structure of the network of layers imposes itself more strongly on the eigenvalues of the entire interconnected structure. This comes as a consequence of the interlacing theorems shown in Sect. 7.3.3 on the main text.

Our findings can be related to the structural transition shown in [63], where the authors evaluated the supra-Laplacian matrix as a function of the inter-layer weights. Their main result is an abrupt structural transition from a decoupled regime, where the layers seem to be independent, to a coupled regime where the layers behave as one single system. Here, we are interested in the supra-adjacency tensor; however, we found a similar phenomenological behavior and a structural change of the system as a function of the inter-layer weights, which in our case are determined by a dynamical process.

7.3.6.2   Localization Analysis

Figures 7.7 and 7.8 show the IPR(Λ_1). On the main panel we present the individual contribution of each layer, while on the insets we have the total IPR(Λ_1). On Fig. 7.7 on the top panel we have the line $(2.3 + 2.9 + 2.6)$, whereas on the bottom panel we have the multiplex network. Similarly, Fig. 7.8 shows the IPR(Λ_1) of tensor $\mathcal{R}$ for the lines $(2.3 + 2.6 + 2.9)$ and $(2.6 + 2.3 + 2.9)$ on (a) and (b), respectively.

As observed in [24], an interesting phenomenon can be observed comparing the different configurations of the network of layers. The largest eigenvalue of the whole system, Λ_1, has its associated eigenvector localized in the dominant layer, that is, in the layer generated using $\gamma = 2.3$. Regarding the line configuration, depending on the position of that layer in the whole system—i.e., central or peripheral layer—the contribution of the nondominant layers to the IPR(Λ_1) varies. In particular, when the dominant layer corresponds to an extreme node of the network of layers, the contribution of the other two layers will be ordered according to the distance to the dominant one. Consequently, when the dominant layer is in the center of the network of layers, the contributions of the nondominant ones are comparable.

Furthermore, for the first eigenvalue, which is usually enough to analyze the localization as a first order approximation, we observe that the layer with the largest eigenvalue dominates the dynamics. In addition, note the similarities between the multiplex and the line configuration $(2.6 + 2.3 + 2.6)$ (see also Fig. 7.8), where

Fig. 7.7 Contribution of each layer to the inverse participation ratio for the first eigenvalue of $\mathcal{R}(\lambda, \eta)$ considering all three layer configurations as a function of the ratio $\frac{\eta}{\lambda}$. On the inset we show the behavior of IPR(Λ_1). Such eigenvalue is related to the leading eigenvalue of the layer $\gamma = 2.3$ when $\frac{\eta}{\lambda} = 0$. On (**a**) we have the line $(2.3 + 2.9 + 2.6)$, while on (**b**) the multiplex case

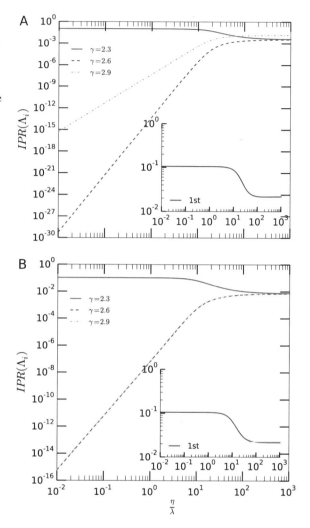

the nondominant layers behave similarly. This is because for small values of $\frac{\eta}{\lambda}$, the effect of the extra edge in the network of layers (closing the triangle) is of second order and thus the similar behavior observed for the two configurations. As $\frac{\eta}{\lambda}$ grows, the symmetry in the node-aligned multiplex dominates the eigenvector structure and the contributions of all layers are comparable.

Finally, Fig. 7.9 shows the tenth larger eigenvalues of the 3-layer multiplex case. The dashed lines represent the leading eigenvalue of each layer. Note that the leading eigenvalue of the layer with $P(k) \sim k^{-2.9}$ is the seventh larger on the network spectrum when $\frac{\eta}{\lambda} = 0$. We observe that there are no crossings on the observed eigenvalues, which is an expected result since the layers have different structures. Furthermore, it is important to remark that all networks of layers evaluated also

Fig. 7.8 Contribution of each layer to the inverse participation ratio for the first eigenvalue of $\mathcal{R}(\lambda, \eta)$ considering all three layer configurations as a function of the ratio $\frac{\eta}{\lambda}$. On the inset we show the behavior of IPR(Λ_1). Such eigenvalue is related to the leading eigenvalue of the layer $\gamma = 2.3$ when $\frac{\eta}{\lambda} = 0$. On (**a**) we have the line $(2.3 + 2.6 + 2.9)$, while on (**b**) the line $(2.6 + 2.3 + 2.9)$

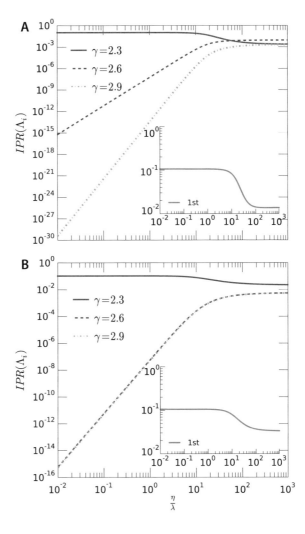

show similar qualitative behaviors. The topology of the network of layers does not lead to qualitative differences on the dependence of Λ_i on $\frac{\eta}{\lambda}$ for the first ten eigenvalues. We also notice that although it is only an approximation, the perturbation theory would be valid roughly up to $\frac{\eta}{\lambda} \lesssim 10$.

It is worth mentioning that the results presented in this section are closely related to an epidemic spreading, as shown in [24]. In such case, the spreading of a disease through different layers can be related to the individual contributions of each layer to the total IPR(Λ_1). Additionally, in the context of disease spreading, the so-called barrier effect was found, where the central layer act as a barrier for the spreading. In this phenomenon the IPR(Λ_1) plays a major role. We believe that we could potentially find similar behaviors for other dynamical processes. Naturally, their physical interpretation would depend on their particular context.

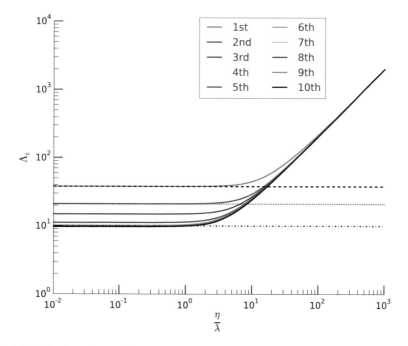

Fig. 7.9 Evaluation of the 8 first eigenvalues of $\mathcal{R}(\lambda, \eta)$ for the multiplex configuration as a function of the ratio $\frac{\eta}{\lambda}$. It is noteworthy that such plot is visually equivalent for all the layer topologies composed by three layers. The dashed lines represent the individual layer leading eigenvalues

7.4   Degree–Degree Correlations in Multilayer Networks

Degree–degree correlations is a fundamental property of single-layer networks, impacting the spreading of diseases, synchronization phenomena, and systems' resilience [4, 55]. Additionally, it has been reported that different correlations arise in different kinds of networks: social networks are in general assortative, meaning that highly connected nodes tend to link with each other, whereas technological and biological systems have disassortative structures, in which high degree nodes are likely attached to low degree nodes [54].

For networks made up of more than one layer, only recently, Nicosia and Latora [56] considered the correlation between the degrees in two different layers. However, their methodology is only for node-aligned multiplex networks, which are special cases of multilayer networks (see [44]). In fact, multiplex networks are made up of n nodes that can be in one or more interacting layers. The links in each layer represent a given mode of interaction between the set of nodes belonging to that layer, whereas links connecting different layers stand for the different modes of interaction between objects involved in [44].

In this section, we study degree–degree correlations in multilayer systems and propose a way to generalize previous assortativity metrics by considering the tensorial formulation introduced in [25]. Our approach also covers a weighted version of assortativity [47] and the case in which the assortativity is given by the Spearman correlation coefficient, generalizing the definition in [48]. The study of a real dataset corresponding to the airport transportation network shows a contrasting behavior between the analyses of each layer independently and altogether, which reinforces the need for such a generalization of the assortativity measure.

7.4.1 Assortativity in Multiplex Networks

Using the tensorial notation, presented in Sect. 7.1 and following Einstein's summation convention, the assortativity coefficient (originally proposed in [54]) can be written as

$$
\rho(\mathcal{W}_\beta^\alpha) = \frac{\mathcal{M}^{-1} \mathcal{W}_\beta^\alpha Q^\beta Q_\alpha - \left[1/2 \mathcal{M}^{-1} \left(\mathcal{W}_\beta^\alpha Q_\alpha u^\beta + \mathcal{W}_\beta^\alpha Q^\beta u_\alpha \right) \right]^2}{\mathcal{M}^{-1} \left(\mathcal{W}_\beta^\alpha (Q_\alpha)^2 u^\beta + \mathcal{W}_\beta^\alpha (Q^\beta)^2 u_\alpha \right) - \left[1/2 \mathcal{M}^{-1} \left(\mathcal{W}_\beta^\alpha Q_\alpha u^\beta + \mathcal{W}_\beta^\alpha Q^\beta u_\alpha \right) \right]^2}
\tag{7.22}
$$

where u is the 1-tensor, which is a tensor of rank 1 and has all elements equal to 1, $\mathcal{W}_\beta^\alpha$ is a second order tensor that summarizes the information that is being extracted and $\mathcal{M} = \mathcal{W}_\beta^\alpha U_\alpha^\beta$ is a normalization constant.

Let us explain in more detail all terms appearing in the expression of $\rho(\mathcal{W}_\beta^\alpha)$. First, we define

$$
Q^\alpha = \mathcal{W}_\beta^\alpha u^\beta,
\tag{7.23}
$$

which is a 1-contravariant tensor and

$$
Q_\beta = \mathcal{W}_\beta^\alpha u_\alpha
\tag{7.24}
$$

which is a 1-covariant tensor. Moreover, the indices are related to the direction of the relationships between nodes. Such a choice ensures a more general expression, capturing degree correlations on nonsymmetric tensors and, consequently, in directed and weighted networks.

Due to the multiplex nature of such systems, we obtain different types of correlations, which can be uncovered by operating on the adjacency tensor. As operations on the nodes we might cite the single layer extraction (Eq. (7.3)), the projected network (Eq. (7.5)), and the overlay network (Eq. (7.6)). Complementary, if we extract the network of layers (Eq. (7.4)), the correlation between different layers can also be evaluated (see Sect. 7.2 and Ref. [25]). In other words, we define the tensor $\mathcal{W}_{\tilde{\delta}}^{\tilde{\gamma}}$ as one of the above-cited projections.

Regarding the overlay and the projected networks, extracting degree–degree correlations, nodes with similar degrees connected in the same or different layers contribute positively to the assortativity coefficient. On the other hand, the connections between hubs and low degree nodes in the same or different layers decrease the assortativity. Self-edges always increase the assortativity, which yields different values of assortativity for the overlay and the projected networks. This gives information on the nature of the coupling between different replicas of the same object among different layers.

Complementary, if we extract the network of layers (Eq. (7.4), Sect. 7.2), the correlation between different layers can also be evaluated. We use $\mathcal{W}^{\tilde{\gamma}}_{\tilde{\delta}} = \Psi^{\tilde{\gamma}}_{\tilde{\delta}}$. It is important to stress that the components of this adjacency tensor are not binary, but weighted by the number of edges inter each layer. Moreover, also in this case, the resulting tensor presents self-edges that encode the information about the density of connections inside a single layer. Finally, we can consider only inter-layer relationships over two different layers. Such information is extracted by projecting the adjacency tensor on the canonical base as

$$\mathcal{W}^{\alpha}_{\beta} = C^{\alpha}_{\beta}(\tilde{r}\tilde{h}) = M^{\alpha\tilde{\delta}}_{\beta\tilde{\gamma}} E^{\tilde{\gamma}}_{\tilde{\delta}}(\tilde{r}\tilde{h}). \qquad (7.25)$$

Note that this is only applicable to multilayer networks and does not make sense in multiplex networks, since in the latter case the coupling is diagonal.

Moreover, in some applications, it is interesting to calculate a pair-wise correlation between a set of nodes, for instance, between couple of layers. Thus, we propose a new operation, that we call selection, which is a projection over a selected set of layers:

$$\mathcal{W}^{\alpha}_{\beta}(\mathcal{L}) = S^{\alpha}_{\beta}(\mathcal{L}) = M^{\alpha\tilde{\delta}}_{\beta\tilde{\gamma}} \Omega^{\tilde{\gamma}}_{\tilde{\delta}}(\mathcal{L}), \qquad (7.26)$$

where $\Omega^{\tilde{\gamma}}_{\tilde{\delta}}$ is a tensor used to select the set of layers we consider in the projection $(\mathcal{L})$. The components of the tensor are equal to unity when the layers $\tilde{\delta}$ and $\tilde{\gamma}$ are selected, and zero otherwise. Note that by selecting all layers together we recover the 1-tensor $U^{\tilde{\gamma}}_{\tilde{\delta}}$ and consequently Eq. (7.5). Another special case is $\tilde{\delta} = \tilde{\gamma}$, which yields Eq. (7.3), or the layer extraction. The tensor can also be generalized to weight different layers. In this case, each element of $\Omega^{\tilde{\gamma}}_{\tilde{\delta}}$ contains the weight of the relationship between two layers $\tilde{\delta}$ and $\tilde{\gamma}$. Such projection is similar to the covariance matrix in statistics, which generalizes the concept of variance. The covariance between two variables is quantified in each entry of the matrix and the main diagonal has the variance of each variable. Thus, we can define a matrix that generalizes the assortativity in a similar way as the covariance matrix generalizes the concept of variance, i.e.

$$\mathbf{S}^{\tilde{\gamma}}_{\tilde{\delta}} = \rho\left(S^{\alpha}_{\beta}(\mathcal{L} = \{\tilde{\gamma}, \tilde{\delta}\})\right), \qquad (7.27)$$

which belongs to a $\mathbb{R}^{L \times L}$ space. We call $\mathbf{S}$ the *P-assortativity matrix*.

Also in this case, a similar operation for the overlay network can be considered, yielding

$$W_\beta^\alpha(\mathcal{L}) = Z_\beta^\alpha(\mathcal{L}) = \sum_{\tilde{h} \in \mathcal{L}}^{L} C_\beta^\alpha(\tilde{h}\tilde{h}), \tag{7.28}$$

which can also be generalized in a similar way as Eqs. (7.26) and (7.27), resulting in the matrix

$$\mathbf{Z}_{\tilde{\delta}}^{\tilde{\gamma}} = \rho \left(Z_\beta^\alpha(\mathcal{L} = \{\tilde{\gamma}, \tilde{\delta}\}) \right). \tag{7.29}$$

We call $\mathbf{Z}$ the *O-assortativity matrix*. A similar inter-layer correlation was also proposed in [56], where the authors suggested measuring the degree correlation between two different layers of the replica of the same object (or node). Furthermore, they proposed three different ways: the Pearson correlation coefficient, Spearman rank correlation, and the Kendall's τ index. However, it is worth pointing out that such an approach does not consider the intra-layer relationship because it is only for node-aligned multiplex networks [44]. Here, we generalize such a measure in terms of tensorial notation.

Finally, the assortativity coefficient can also be defined in terms of the Spearman rank correlation [48], since the traditional definition of this coefficient based on the Pearson correlation [54] can lead to incomplete results, as discussed in [48]. The generalization of assortativity coefficient allows to consider the Spearman rank correlation coefficient by changing Eqs. (7.23) and (7.24). Specifically, instead of considering the values of Q^α and Q_β, one substitutes them by their respective ranks.[1] Such transformation is performed by using

$$Q^\alpha = \mathrm{rank}(W_\beta^\alpha u^\beta) \tag{7.30}$$

and

$$Q_\beta = \mathrm{rank}(W_\beta^\alpha u_\alpha), \tag{7.31}$$

where $\mathrm{rank}(X_i)$ is the rank of the tensor X_i.

We henceforth denote by $\rho^P(W_\beta^\alpha)$ and $\rho^S(W_\beta^\alpha)$ the Pearson and Spearman correlation coefficients, respectively. Furthermore, we adopt $(\mathbf{S}^P)_{\tilde{\delta}}^{\tilde{\gamma}}$ and $(\mathbf{S}^S)_{\tilde{\delta}}^{\tilde{\gamma}}$ for the pair-wise correlation matrices using the Pearson and Spearman correlation coefficients, respectively. The same notation can be used for the matrices $(\mathbf{Z}^P)_{\tilde{\delta}}^{\tilde{\gamma}}$

---

[1]One may not confuse rank in this context with the tensorial rank. Here it is the position in the ordered set of values, whereas the rank of a tensor is the number of covariant and contravariant indices.

and $(\mathbf{Z}^S)^{\tilde{\gamma}}_{\tilde{\delta}}$. Monoplex assortativity, i.e., assortativity in single-layer networks [54], is recovered by considering the adjacency matrix, $\mathcal{W}^\alpha_\beta = A^\alpha_\beta$, and consequently Q^α and Q_β are analogous to in-degree and out-degree, respectively. Note that $Q^\alpha = Q_\beta$ for undirected networks. Moreover, $\mathcal{M}$ is equal to twice the number of edges, recovering the equation introduced in [54], which also captures correlations of weighted networks, as exposed in [47].

Each approach presented here gives a different descriptor of the multilayer structure. For instance, the projected and overlay networks gather the information of all layers into a single layer structure, aiming at describing the whole system using single descriptors. Aside from those, we also provide a pair-wise descriptor which gives another type of information. Besides, there is also the network of layers, which possesses information about yet another level of the system. In this way, approach gives a set of metrics that capture information about the whole multilayer structure. However, it is worth mentioning that the interpretation and choices depend on the application.

7.4.2 Application to Real Data

We analyze the airport transportation network [16], whose multilayer representation was studied in [13]. The network comprises 450 airports and 37 companies, which are mapped as nodes and layers, respectively. More specifically, in each layer, the edges represent the directed flights operated by a given company and nodes, airports. Figure 7.10 shows a representation of 12 layers of such multiplex network. The inter-layer connections link the airports shared by pairs of different companies. This approach gives us a multiplex network that is not a node aligned multiplex network since the latter considers a diagonal coupling between all nodes in all layers. Note that the way proposed in [13] to create the aggregated monoplex network is the union of all layers considering multiple edges as single ones. This is in contrast to our approach, because we consider the projections and overlay

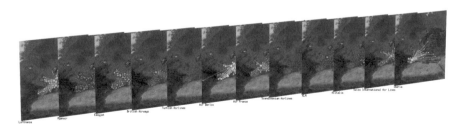

Fig. 7.10 Example of an airport transportation multiplex network. Each layer represents an airline, in which each node represents an airport and the edges are flights between two airports. This visualization was generated using MuxViz [26]

Table 7.2 Structural properties of the airport transportation multiplex networks

Network	N	$\mathcal{M}$	$\langle Q^\alpha \rangle$	$\rho^P(W_\beta^\alpha)$	$\rho^S(W_\beta^\alpha)$
Network of layers (Ψ_β^α)	37	30398.0	821.568	0.377	0.286
Overlay (O_β^α)	450	7176.0	15.947	−0.050	−0.025
Projected network (P_β^α)	450	30398.0	67.551	0.795	0.560

networks as weighted networks, thus retaining the information of the number of different connections between the same pair of airports.

Previous studies [13, 16] showed that the airport transportation network presents the rich-club effect, which refers to the tendency of highly central nodes to be connected among themselves. This is also captured by the assortativity as shown in Table 7.2, where we verify that the projected network has positive assortativity coefficients, agreeing with previous analyses. However, note that the projection has a positive value of the assortativity, whereas the overlay has a negative one. Thus, the assortativity of the projection indicates that many companies share hubs airport, not that hubs connect between them. This apparent contradiction results from the fact that the rich-club effect is masked out in the overlay setup by a large number of peripheral nodes connecting to hubs.

The analysis of each layer separately (see Table 7.3) shows a different result, where most of the layers are disassortative. The only exception is the Netjet layer, which presents a positive coefficient for the rank correlation. Usually, the companies focus their activities in one city or country, for example, Lufthansa in Germany or Air France in France, and have flights to other airports where their activity is lower. This leads to the disassortative behavior of each layer. Additionally, the disassortative correlations found in single layers is more pronounced than that of the overlay representation, which can be explained by noticing that hubs of a company are peripheral (or secondary) airports for other companies, but when the layers are collapsed they are also hubs in the overlay network and are connected.

Figures 7.11 and 7.12 show the pair-wise correlation between layers. Interestingly, the latter is disassortative, in contrast to the results obtained for the projected network, but of the same sign as those computed for the overlay representation (see Table 7.2). Furthermore, our construction of the adjacency tensor leads to an assortative network of layers, suggesting that bigger companies tend to share similar airports. This analysis agrees with [13], where the authors argued that the main airports are connected to each other via directed flights. In addition, considering the Pearson correlations, the *O-assortativity* matrix presents lower values if compared to the *P-assortativity* matrix due to the intra-layer contributions, as discussed before.

Table 7.3 Structural properties of each layer of the airport transportation multiplex networks

Company	N	$\mathcal{M}$	$\langle Q^\alpha \rangle$	$\rho^P(C_\beta^\alpha)$	$\rho^S(C_\beta^\alpha)$
Lufthansa	106	488.0	4.604	−0.668	−0.473
Ryanair	128	1202.0	9.391	−0.321	−0.348
Easyjet	99	614.0	6.202	−0.428	−0.470
British Airways	65	132.0	2.031	−0.775	−0.754
Turkish Airlines	86	236.0	2.744	−0.697	−0.567
Air Berlin	75	368.0	4.907	−0.501	−0.434
Air France	59	138.0	2.339	−0.637	−0.661
Scandinavian Airlines	66	220.0	3.333	−0.681	−0.521
KLM	63	124.0	1.968	−1.000	−1.000
Alitalia	51	186.0	3.647	−0.572	−0.538
Swiss International Air Lines	48	120.0	2.500	−0.728	−0.618
Iberia	35	70.0	2.000	−0.900	−0.838
Norwegian Air Shuttle	52	174.0	3.346	−0.511	−0.523
Austrian Airlines	67	144.0	2.149	−0.823	−0.744
Flybe	43	198.0	4.605	−0.560	−0.489
Wizz Air	45	184.0	4.089	−0.350	−0.381
TAP Portugal	42	106.0	2.524	−0.779	−0.610
Brussels Airlines	44	86.0	1.955	−1.000	−1.000
Finnair	42	84.0	2.000	−0.915	−0.858
LOT Polish Airlines	44	110.0	2.500	−0.658	−0.598
Vueling Airlines	36	126.0	3.500	−0.438	−0.456
Air Nostrum	48	138.0	2.875	−0.571	−0.569
Air Lingus	45	116.0	2.578	−0.670	−0.625
Germanwings	44	134.0	3.045	−0.628	−0.482
Panagra Airways	45	116.0	2.578	−0.625	−0.593
Netjets	94	360.0	3.830	−0.106	0.107
Transavia Holland	40	114.0	2.850	−0.585	−0.535
Niki	36	74.0	2.056	−0.838	−0.784
SunExpress	38	134.0	3.526	−0.797	−0.542
Aegean Airlines	38	106.0	2.789	−0.583	−0.560
Czech Airlines	42	82.0	1.952	−1.000	−1.000
European Air Transport	53	146.0	2.755	−0.416	−0.423
Malev Hungarian Airlines	35	68.0	1.943	−1.000	−1.000
Air Baltic	45	90.0	2.000	−0.844	−0.812
Wideroe	45	180.0	4.000	−0.293	−0.311
TNT Airways	53	122.0	2.302	−0.415	−0.346
Olympic Air	37	86.0	2.324	−0.754	−0.662

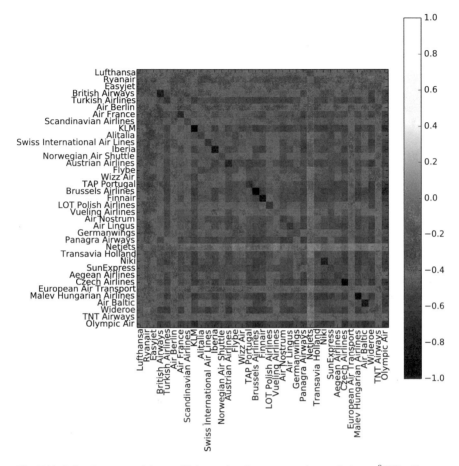

Fig. 7.11 Pair-wise assortativity coefficient using Spearman rank correlation, $\rho^S(\mathbf{S}^\alpha_\beta)$. Observe that the main diagonal presents the same coefficient considering the layer extraction operation, $\rho^S(C^\alpha_\beta(\tilde{r}\tilde{r}))$

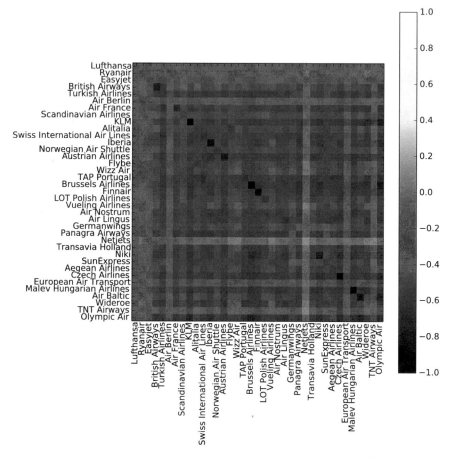

Fig. 7.12 Pair-wise assortativity coefficient using Spearman rank correlation, $\rho^S(\mathbf{Z}_\beta^\alpha)$. Observe that the main diagonal presents the same coefficient considering the layer extraction operation, $\rho^S(C_\beta^\alpha(\tilde{r}\tilde{r}))$

Conclusions

As mentioned in the Introduction to this Springer Brief, the main goal of this text is to further motivate research in the topic of multiplex/multilayer networks. By presenting the main concepts of the multilayer framework, we hope that we have provided the reader with a set of tools and concepts that will eventually allow using the formalism and gaining a stronger formal basis. We believe that choosing the matricial or the tensorial notation is a matter of taste and convenience. For some problems, one might be easier or more natural than the other, since each one emphasizes a different set of properties of our system. Additionally, knowing both also allows the researcher to make use of their specific advantages.

We would also like to highlight some important and interesting research perspectives in this area. For instance, in Chap. 3, we formally defined a small set of properties needed to properly define a structural metric. Based on these concepts, we believe that many measures and metrics can be extended from monoplex networks to multilayer ones. In doing that, we must be careful when interpreting the results as well as when doing some approximations as both have been shown to depend on the dynamics and the problem been tackled more often than not.

Regarding the spectral properties of multilayer systems, we also stress that the richness of this kind of system is directly reflected on its spectra. In this Springer Brief, we mainly used perturbation theory and interlacing properties of quotient graphs in Chap. 4 and the new polynomial eigenvalue interpretation in Chap. 6. With these tools, we were able to extract many structural and even dynamical insights for processes occurring on top of multilayer networks. However, new analytical and numerical tools and techniques are needed for the cases in which the multilayer structure is more complex, which will surely allow exploring more realistic systems.

Finally, we also reiterate that this book has an introductory character, presenting the main formalism that several authors have developed in the last few years. In this sense, this text is not exhaustive as we have not presented alternative approaches. Likewise, we have focused on formal and structural aspects of multilayer networks, leaving aside the already vast literature dealing with dynamical processes

© The Author(s) 2018
E. Cozzo et al., *Multiplex Networks*, SpringerBriefs in Complexity,
https://doi.org/10.1007/978-3-319-92255-3

in multilayer systems. We however hope that the reader has grasped the basic formalism and aspects of multilayer systems so as she/he can develop her/his own research agenda in this fascinating area.

Symbols and Notation

Matricial Notation

- $G(V, E)$: a graph, where V is a set of nodes and E is a set of edges
- $(u, v) \in E$ with $u, v \in V$: and edge of a graph $G(V, E)$; u and v are said to be *adjacent*
- $u \sim v$: adjacency relation
- $\mathcal{M} = (V, L, P, M)$: ordered quadruple representing a multiplex network
- $L = \{1, \ldots, m\}$: layer set of a multiplex network; a layer $\alpha \in L$ is an index that represents a particular type of interaction or relation
- $\mid L \mid = m$: the number of layers in a multiplex network
- $G_P = (V, L, P)$, where $P \subseteq V \times L$: a binary relation that indicates which node of a multiplex participates in which layers. $G_P = (V, L, P)$ can be interpreted as a (bipartite) graph where P is the edge set
- $(u, \alpha) \in P$: an ordered tuple called node-layer pair; it is the representative of node u in layer α
- P: the set of the node-layer pairs
- $\mid V_\beta \mid = n_\beta$: the number of node-layer pairs in layer β
- *node-aligned multiplex*: a multiplex network in which each node $u \in V$ has a representative in each layer $\alpha \in L$, i.e., $P = V \times L$
- $G_\beta(V_\beta, E_\beta)$, where $V_\beta = \{(u, \alpha) \in P \mid \alpha = \beta\}$: a *layer-graph*
- $M = \{G_\alpha\}_{\alpha \in L}$: the set of all layer-graph
- $G_C(P, E_C)$: the coupling graph in which there is an edge between two node-layer pairs (u, α) and (v, β) if and only if $u = v$
- *supra-nodes*: the $n = \mid P \mid$ disconnected components which formed G_C. Each component is formed by all the representatives of a node in different layers
- $G_l = \bigcup_\alpha G_\alpha$: the *intra-layer graph*. The union of all the layer-graphs, i.e., $G_l = \bigcup_\alpha G_\alpha$
- $G_\mathcal{M} = G_l \cup G_C$: the *supra-graph*

© The Author(s) 2018
E. Cozzo et al., *Multiplex Networks*, SpringerBriefs in Complexity,
https://doi.org/10.1007/978-3-319-92255-3

- $l(u) = (u, \alpha) \in P \mid \alpha \in L$: the set of node-layer pairs that correspond to the same node u
- $l^{-1}(i)$: the unique node that corresponds to the node-layer pair i
- $\mathbf{A}(G)$: the adjacency matrix of the graph $G(V, E)$
- $\mathbf{L}(G)$: Laplacian matrix of the graph $G(V, E)$
- $\mathbf{K}_n$: the adjacency matrix of a complete graph on n nodes
- $\mathbf{A}^{(\alpha)}$: layer adjacency matrix; the adjacency matrix of the layer-graph G_α
- $\mathbf{L}^{(\alpha)}$: layer-Laplacian, the Laplacian matrix of the layer-graph G_α
- $\mathbf{P}$: participation matrix; the adjacency matrix of the participation graph G_P
- $\mathcal{C}$: the coupling matrix; the adjacency matrix of the coupling graph G_C
- $\mathcal{L}_C$: the Laplacian of the coupling graph
- *standard labeling*: a labeling of the node-layer pairs such that the coupling matrix results in a block matrix whose diagonal blocks are all zeros and such that the same row and column in different layer adjacency matrices correspond to the representative of the same node in different layers
- $\bar{A}$: the *supra-adjacency* matrix; the adjacency matrix of the supra-graph $G_\mathcal{M}$. Assuming the standard labeling $\bar{A} = \bigoplus_\alpha \mathbf{A}^\alpha + \mathcal{C}$
- $A = \bigoplus \mathbf{A}^\alpha$: intra-layer adjacency matrix; the adjacency matrix of the intra-layer graph G_l
- $\bar{\mathcal{L}} = L(G_\mathcal{M})$: the *supra-Laplacian* matrix; the Laplacian matrix of the supra-graph $G_\mathcal{M}$. Assuming the standard labeling $\bar{\mathcal{L}} = \bigoplus_\alpha \mathcal{L}^{(\alpha)} + \mathcal{L}_C$
- K_i: the degree of a node-layer i; the number of node-layers connected to it by an edge in $G_\mathcal{M}$
- $e_\alpha = \sum_{\beta < \alpha} n_\beta$: excess index of layer α
- $k_{i(\alpha)}$: layer-degree of a node-layer i; the number of neighbors it has in G^α
- $c_{i(\alpha)}$: the coupling degree of a node-layer i; the number of neighbors it has in G_C
- $\widehat{C}(\beta, \gamma) = \beta \mathcal{I} + \gamma \mathcal{C}$
- *multiplex walk matrix*: a matrix that encodes the permissible steps in a multiplex network
- $\mathcal{A}\widehat{C}$: multiplex walk matrix that encodes the steps in which after each intra-layer step a walk can continue on the same layer
- $\widehat{C}\mathcal{A}$: multiplex walk matrix that encodes the steps in which before each intra-layer step a walk can continue on the same layer
- *elementary cycle*: a term that consists of products of the matrices $\mathcal{A}$ and $\mathcal{C}$ (i.e., there are no sums) after one expands the expression for a cycle (which is a weighted sum of such terms)
- *auxiliary supra-graph*: the graph defined by a multiplex walk matrix when interpreted as an adjacency matrix
- *dimensionality reduction*: an operation that aggregate the interaction pattern of different layers
- *aggregate network*: the result of a dimensionality reduction of a multiplex network
- *quotient graph* $\mathcal{Q}(G)$: a coarsening of a graph G with respect to a partition $\{V_1, V_2, \ldots, V_n\}$. It has one node per cluster V_i, and an edge from V_i to V_j weighted by an average connectivity from V_i to V_j (see Sect. 2.3.1 for the definitions of left, right, and symmetric quotient)

- *equitable partition*: a partition of the node set of a graph such that the number of edges (taking weights into account) from a node in V_i to any node in V_j is independent of the chosen node in V_i
- *regular quotient*: the quotient graph associated with an equitable partition
- *almost equitable partition*: a partition for which the regularity condition is satisfied for all $i \neq j$ (but not necessarily for $i = j$)
- *almost regular quotient*: the quotient associated with an almost equitable partition
- $\Lambda = diag\{\kappa_1, \ldots, \kappa_n\}$: the multiplexity degree matrix
- S_n: the node partition characteristic matrix; $s_{iu} = 1$ if and only if the node-layer i is a representative of node u
- S_l: the layer characteristic matrix; $s_{i\alpha} = 1$ only if the node-layer i is in layer α
- $\tilde{A} = \Lambda^{-1} S_n^T \bar{A} S_n$: the adjacency matrix of the aggregate network
- $\tilde{W} = \tilde{A} - diag(\tilde{A})$: the adjacency matrix of the loop-less aggregate network
- $W = S_n^T A S_n$,: the adjacency matrix of the sum aggregate network
- $\tilde{A}_l$: the adjacency matrix of the network of layers
- local clustering coefficient C_u: in an unweighted single-layer network it is the number of triangles (i.e., triads) that include node u divided by the number of connected triples with node u in the center
- *global clustering coefficient*: the ratio between the mean number of closed triple and the mean number of open triples
- $c_{*,i} = \frac{t_{*,i}}{d_{*,i}}$: local clustering coefficient of the node-layer i
- $C_{*,u} = \frac{\sum_{i \in l(u)} t_{*,i}}{\sum_{i \in l(u)} d_{*,i}}$: local clustering coefficient of the node u
- $C_* = \frac{\sum_i t_{*,i}}{\sum_i d_{*,i}}$: global clustering coefficient
- *f-centrality* of a node u in a single-layer network: defined as $f(\mathbf{A})_{uu}$
- *sub-graph centrality* G_u of a node in a single layer network: is given by $exp(\mathbf{A}))_{uu}$
- *f-communicability* between two distinct nodes u and v in a single-layer network: defined as $f(\mathbf{A})_{uv}$
- *communicability matrix*: $\mathbf{G} = \exp(\mathbf{A})$
- *f-centrality* of a node-layer pair i: $k_i = c_1 f(\mathcal{A}\hat{\mathcal{C}})_{ii} + c_2 f(\hat{\mathcal{C}}\mathcal{A})_{ii}, c_1, c_2 > 0$
- *f-communicability* between two distinct node-layer pairs i and j: $k_{ij} = c_1 f(\mathcal{A}\hat{\mathcal{C}})_{ij} + c_2 f(\hat{\mathcal{C}}\mathcal{A})_{ij}, c_1, c_2 > 0$
- *supra-communicability matrix*: $\mathcal{K} = c_1 f(\mathcal{A}\hat{\mathcal{C}}) + c_2 f(\mathcal{A}\hat{\mathcal{C}})^T$
- *aggregate f-communicability matrix*: $\tilde{\mathbf{K}} = c_1 \mathcal{Q}_R(f(\mathcal{A}\hat{\mathcal{C}})) + c_2 \mathcal{Q}_R(f(\mathcal{A}\hat{\mathcal{C}}))^T$
- *spectral radius*: the largest eigenvalue of the adjacency matrix of a graph
- *algebraic connectivity*: the second-smallest eigenvalue of the Laplacian of a graph
- *dominant layer*: the layer with the largest spectral radius in a multiplex network
- *Laplacian dominant layer*: the layer with the lowest algebraic connectivity in a multiplex network
- *effective multiplexity* z: the weighted mean of the coupling degree with the weight given by the squares of the entries of the leading eigenvector of $\mathcal{A}$
- *correlated multiplexity* $z_c = \frac{\phi^T \mathcal{C} \phi}{\phi^T \phi}$, where ϕ is the eigenvector associated with the largest eigenvalue of $\mathcal{A}$

- *Aggregate-Equivalent Multiplex (AEM)*: the Cartesian product between the aggregate network and the network of layers of a given multiplex network
- $\|\|\cdot\|\|$ is a matrix metric
- $\|\mathbf{A}\|_2$ is the spectral norm of the matrix $\mathbf{A}$

Tensorial Notation

- $M \in \mathbb{R}^{n \times n \times m \times m}$: the fourth-order adjacency tensor of a multilayer network with m layers and n nodes on each layer
- $\mathcal{R}^{\alpha\tilde{\gamma}}_{\beta\tilde{\delta}}(\lambda, \eta)$: supra contact tensor
- $A^{\alpha}_{\beta}(\tilde{r})$: the second order tensor for the layer $\tilde{r}$
- $u_{\alpha} \in \mathbb{R}^n$: all one tensor
- $U^{\beta\tilde{\delta}} \in \mathbb{R}^{n \times m}$: all one tensor
- $\Psi^{\tilde{\gamma}}_{\tilde{\delta}} \in \mathbb{R}^{m \times m}$: network of layers
- $P^{\alpha}_{\beta} \in \mathbb{R}^{n \times n}$: projection or projected network
- $O^{\alpha}_{\beta} \in \mathbb{R}^{n \times n}$: overlay network
- Λ_i: Eigenvalue i. Note that on the tensorial notation we ordered as $\Lambda_1 \geq \Lambda_2 \geq \ldots \geq \Lambda_{nm}$
- Λ^l_i: individual layer eigenvalue i associated with the layer l
- IPR(Λ): Inverse participation ratio of the eigentensor (also applies for an eigenvector) associated with the eigenvalue Λ
- $\Phi^{\tilde{\gamma}}_{\tilde{\delta}}(\lambda, \eta)$: normalized network of layers
- $\mathbf{P}^{\alpha}_{\beta}$: normalized projection
- $\rho(\mathcal{W}^{\alpha}_{\beta})$: Assortativity coefficient of the graph/projection represented by the tensor $\mathcal{W}^{\alpha}_{\beta}$
- $\rho^P(\mathcal{W}^{\alpha}_{\beta})$: the Pearson correlation coefficient of the graph/projection represented by the tensor $\mathcal{W}^{\alpha}_{\beta}$
- $\rho^S(\mathcal{W}^{\alpha}_{\beta})$: the Spearman correlation coefficients of the graph/projection represented by the tensor $\mathcal{W}^{\alpha}_{\beta}$
- $\mathbf{S}^{\tilde{\gamma}}_{\tilde{\delta}} \in \mathbb{R}^{L \times L}$: *P-assortativity matrix*
- $\mathbf{Z}^{\tilde{\gamma}}_{\tilde{\delta}} \in \mathbb{R}^{L \times L}$: *O-assortativity matrix*

Bibliography

1. S.E. Ahnert, D. Garlaschelli, T.M.A. Fink, G. Caldarelli, Ensemble approach to the analysis of weighted networks. Phys. Rev. E **76**(1), 016101 (2007)
2. J.A. Almendral, A. Díaz-Guilera, Dynamical and spectral properties of complex networks. New J. Phys. **9**(6), 187 (2007)
3. A. Arenas, A. Díaz-Guilera, C.J. Pérez-Vicente, Synchronization reveals topological scales in complex networks. Phys. Rev. Lett. **96**(11), 114102 (2006)

4. A. Barrat, M. Barthlemy, A. Vespignani, *Dynamical Processes on Complex Networks* (Cambridge University Press, New York, 2008)
5. L. Barrett, S.P. Henzi, D. Lusseau, Taking sociality seriously: the structure of multidimensional social networks as a source of information for individuals. Philos. Trans. R. Soc., B: Biol. Sci. **367**(1599), 2108–2118 (2012)
6. F. Battiston, V. Nicosia, V. Latora, Metrics for the analysis of multiplex networks. Phys. Rev. E **89**, 032804 (2013)
7. J.D. Bolter, R. Grusin, *Remediation: Understanding New Media* (MIT Press, Cambridge, 1999)
8. R.L. Breiger, P.E. Pattison, Cumulated social roles: the duality of persons and their algebras. Soc. Networks **8**(3), 215–256 (1986)
9. P. Bródka, K. Musial, P. Kazienko, A method for group extraction in complex social networks, in *Knowledge Management, Information Systems, E-Learning, and Sustainability Research* (Springer, Berlin, 2010), pp. 238–247
10. P. Bródka, P. Kazienko, K. Musiał, K. Skibicki, Analysis of neighbourhoods in multi-layered dynamic social networks. Int. J. Comput. Intell. Syst. **5**(3), 582–596 (2012)
11. A.E. Brouwer, W.H. Haemers, *Spectra of Graphs* (Springer, New York, 2012)
12. M. Buiatti, M. Buiatti, The living state of matter, in *Rivista di biologia biology forum*, vol. 94 (ANICIA SRL, Roma, 2001), pp. 59–82
13. A. Cardillo, J. Gómez-Gardenes, M. Zanin, M. Romance, D. Papo, F. del Pozo, S. Boccaletti, Emergence of network features from multiplexity. Sci. Rep. **3** (2013)
14. S.E. Chang, H.A. Seligson, R.T. Eguchi, *Estimation of the Economic Impact of Multiple Lifeline Disruption: Memphis Light, Gas and Water Division Case Study* (National Center for Earthquake Engineering Research, Buffalo, 1996)
15. M.T. Chu, Inverse eigenvalue problems. SIAM Rev. **40**(1), 1–39 (1998)
16. V. Colizza, A. Flammini, M.A. Serrano, A. Vespignani, Detecting rich-club ordering in complex networks. Nat. Phys. **2** 110–115 (2006)
17. E. Cozzo, Y. Moreno, Characterization of multiple topological scales in multiplex networks through supra-Laplacian eigengaps. Phys. Rev. E **94**, 052318 (2016)
18. E. Cozzo, R.A. Banos, S. Meloni, Y. Moreno, Contact-based social contagion in multiplex networks. Phys. Rev. E **88**(5) (2013)
19. E. Cozzo, M. Kivela, M. De Domenico, A. Solé-Ribalta, A. Arenas, S. Gómez, M.A. Porter, Y. Moreno, Structure of triadic relations in multiplex networks. New J. Phys. **17**(7), 073029 (2015)
20. E. Cozzo, G.F. Arruda, F.A. Rodrigues, Y. Moreno, Multilayer networks: metrics and spectral properties, in *Interconnected Networks* (Springer, Cham, 2016), pp. 17–35
21. R. Criado, J. Flores, A. García del Amo, J. Gómez-Gardeñes, M. Romance, A mathematical model for networks with structures in the mesoscale. Int. J. Comput. Math. **89**(3), 291–309 (2012)
22. L. da F. Costa, F.A. Rodrigues, G. Travieso, P. Ribeiro, V. Boas, Characterization of complex networks: a survey of measurements. Adv. Phys. **56**(1), 167–242 (2007)
23. K.C. Das, R.B. Bapat, A sharp upper bound on the largest Laplacian eigenvalue of weighted graphs. Linear Algebra Appl. **409**, 153–165 (2005)
24. G.F. de Arruda, E. Cozzo, T.P. Peixoto, F.A. Rodrigues, Y. Moreno, Disease localization in multilayer networks. Phys. Rev. X **7**, 011014 (2017)
25. M. De Domenico, A. Solé-Ribalta, E. Cozzo, M. Kivelä, Y. Moreno, M.A. Porter, S. Gómez, A. Arenas, Mathematical formulation of multilayer networks. Phys. Rev. X **3**(4), 041022 (2013)
26. M. De Domenico, M.A. Porter, A. Arenas, MuxViz: a tool for multilayer analysis and visualization of networks. J. Complex Networks **3**, 159–176 (2015)
27. M. Dickison, S. Havlin, H.E. Stanley, Epidemics on interconnected networks. Phys. Rev. E **85**(6), 066109 (2012)
28. S.N. Dorogovtsev, A.V. Goltsev, J.F.F. Mendes, Critical phenomena in complex networks. Rev. Mod. Phys. **80**(4), 1275 (2008)
29. E. Estrada, J.A. Rodriguez-Velazquez, Subgraph centrality in complex networks. Phys. Rev. E **71**(5), 056103 (2005)

30. L.C. Freeman, Centrality in social networks conceptual clarification. Soc. Networks **1**(3), 215–239 (1979)
31. R. Gallotti, M. Barthelemy, Anatomy and efficiency of urban multimodal mobility. Sci. Rep. **4** (2014)
32. M. Gluckman, C.D. Forde, *Essays on the Ritual of Social Relations* (Manchester University Press, Manchester, 1962)
33. I. Gohberg, P. Lancaster, L. Rodman, *Matrix Polynomials*, vol. 58 (SIAM, Philadelphia, 1982)
34. A.V. Goltsev, S.N. Dorogovtsev, J.G. Oliveira, J.F.F. Mendes, Localization and spreading of diseases in complex networks. Phys. Rev. Lett. **109**, 128702 (2012)
35. S. Gomez, A. Diaz-Guilera, J. Gomez-Gardeñes, C.J. Perez-Vicente, Y. Moreno, A. Arenas, Diffusion dynamics on multiplex networks. Phys. Rev. Lett. **110**(2), 028701 (2013)
36. P. Grindrod, Range-dependent random graphs and their application to modeling large small-world proteome datasets. Phys. Rev. E **66**(6), 066702 (2002)
37. W.H. Haemers, Interlacing eigenvalues and graphs. Linear Algebra Appl. **226**, 593–616 (1995)
38. R.A. Horn, C.R. Johnson, *Matrix Analysis*, 2nd edn. (Cambridge University Press, New York, 2012)
39. A. Jamakovic, P. Van Mieghem, On the robustness of complex networks by using the algebraic connectivity, in *Networking*. Lecture Notes in Computer Science, vol. 4982 (Springer, Berlin, 2008), pp. 183–194
40. B. Kapferer, Norms and the manipulation of relationships in a work context. *Social Networks in Urban Situations* (Manchester University Press, Manchester, 1969)
41. B. Kapferer, *Strategy and Transaction in an African Factory: African Workers and Indian Management in a Zambian Town* (Manchester University Press, Manchester, 1972)
42. M. Karlberg, Testing transitivity in graphs. Soc. Networks **19**(4), 325–343 (1997)
43. S.A. Kauffman, Metabolic stability and epigenesis in randomly constructed genetic nets. J. Theor. Biol. **22**(3), 437–467 (1969)
44. M. Kivela, A. Arenas, M. Barthelemy, J.P. Gleeson, Y. Moreno, M.A. Porter, Multilayer networks. J. Complex Networks **2**(3), 203–271 (2014)
45. D. Krackhardt, Cognitive social structures. Soc. Networks **9**(2), 109–134 (1987)
46. P. Lancaster, *Lambda-Matrices and Vibrating Systems* (Pergamon Press, Oxford, 1966)
47. C.C. Leung, H.F. Chau, Weighted assortative and disassortative networks model. Phys. A: Stat. Mech. Appl. **378**(2), 591–602 (2007)
48. N. Litvak, R. van der Hofstad, Uncovering disassortativity in large scale-free networks. Phys. Rev. E **87**, 022801 (2013)
49. R.D. Luce, A.D. Perry, A method of matrix analysis of group structure. Psychometrika **14**(2), 95–116 (1949)
50. B.D. MacArthur, R.J. Sánchez-García, Spectral characteristics of network redundancy. Phys. Rev. E **80**(2), 026117 (2009)
51. B.D. MacArthur, R.J. Sánchez-García, J.W. Anderson, Symmetry in complex networks. Discret. Appl. Math. **156**(18), 3525–3531 (2008)
52. J. Martín-Hernández, H. Wang, P. Van Mieghem, G. D'Agostino, Algebraic connectivity of interdependent networks. Phys. A: Stat. Mech. Appl. **404**, 92–105 (2014)
53. A. Milanese, J. Sun, T. Nishikawa, Approximating spectral impact of structural perturbations in large networks. Phys. Rev. E **81**, 046112 (2010)
54. M.E.J. Newman, Assortative mixing in networks. Phys. Rev. Lett. **89**, 208701 (2002)
55. M. Newman, *Networks: An Introduction* (Oxford University Press, Oxford, 2010)
56. V. Nicosia, V. Latora, Measuring and modeling correlations in multiplex networks. Phys. Rev. E **92**, 032805 (2015)
57. openflights.org. http://openflights.org/data.htm
58. T. Opsahl, P. Panzarasa, Clustering in weighted networks. Soc. Networks **31**(2), 155–163 (2009)
59. C. Orsini, M.M. Dankulov, P. Colomer-de Simón, A. Jamakovic, P. Mahadevan, A. Vahdat, K.E. Bassler, Z. Toroczkai, M. Boguñá, G. Caldarelli et al., Quantifying randomness in real networks. Nat. Commun. **6** (2015)

60. R. Parshani, S.V. Buldyrev, S. Havlin, Interdependent networks: reducing the coupling strength leads to a change from a first to second order percolation transition. Phys. Rev. Lett. **105**(4), 048701 (2010)
61. F. Passerini, S. Severini, Quantifying complexity in networks: the von Neumann entropy. Int. J. Agent Technol. Syst. **1**(4), 58–67 (2009)
62. F. Radicchi, Driving interconnected networks to supercriticality. Phys. Rev. X **4**(2), 021014 (2014)
63. F. Radicchi, A. Arenas, Abrupt transition in the structural formation of interconnected networks. Nat. Phys. **9**, 717–720 (2013)
64. H. Rainie, B. Wellman, *Networked: The New Social Operating System* (MIT Press, Cambridge, 2012)
65. F.J. Roethlisberger, W.J. Dickson, *Management and the Worker* (Harvard University Press, Cambridge, 1939)
66. M.P. Rombach, M.A. Porter, J.H. Fowler, P.J. Mucha, Core-periphery structure in networks. SIAM J. Appl. Math. **74**(1), 167–190 (2014)
67. M. Saerens, F. Fouss, L. Yen, P. Dupont, The principal components analysis of a graph, and its relationships to spectral clustering, in *Machine Learning: ECML 2004* (2004), pp. 371–383
68. F.D. Sahneh, C. Scoglio, P. Van Mieghem, Exact coupling threshold for structural transition reveals diversified behaviors in interconnected networks. Phys. Rev. E **92**, 040801 (2015)
69. R.J. Sánchez-García, E. Cozzo, Y. Moreno, Dimensionality reduction and spectral properties of multilayer networks. Phys. Rev. E **89**, 052815 (2014)
70. J. Saramäki, M. Kivelä, J.-P. Onnela, K. Kaski, J. Kertesz, Generalizations of the clustering coefficient to weighted complex networks. Phys. Rev. E **75**(2), 027105 (2007)
71. D.J. Selkoe, Alzheimer's disease is a synaptic failure. Science **298**(5594), 789–791 (2002)
72. H.-W. Shen, X.-Q. Cheng, B.-X. Fang, Covariance, correlation matrix, and the multiscale community structure of networks. Phys. Rev. E **82**(1), 016114 (2010)
73. A. Sole-Ribalta, M. De Domenico, N.E. Kouvaris, A. Diaz-Guilera, S. Gomez, A. Arenas, Spectral properties of the Laplacian of multiplex networks. Phys. Rev. E **88**(3), 032807 (2013)
74. S.-W. Son, G. Bizhani, C. Christensen, P. Grassberger, M. Paczuski, Percolation theory on interdependent networks based on epidemic spreading. EPL (Europhys. Lett.) **97**(1), 16006 (2012)
75. A. Sydney, C. Scoglio, D. Gruenbacher, Optimizing algebraic connectivity by edge rewiring. Appl. Math. Comput. **219**(10), 5465–5479 (2013)
76. M. Szell, R. Lambiotte, S. Thurner, Multirelational organization of large-scale social networks in an online world. Proc. Natl. Acad. Sci. **107**(31), 13636–13641 (2010)
77. F. Tisseur, K. Meerbergen, The quadratic eigenvalue problem. SIAM Rev. **43**(2), 235–286 (2001)
78. L.M. Verbrugge, Multiplexity in adult friendships. Soc. Forces **57**(4), 1286–1309 (1979)
79. S. Wasserman, K. Faust, *Social Network Analysis: Methods and Applications*, vol. 8 (Cambridge University Press, Cambridge, 1994)
80. D.J. Watts, S.H. Strogatz, Collective dynamics of 'small-world' networks. Nature **393**(6684), 440–442 (1998)
81. B. Wellman, Physical place and cyberplace: the rise of personalized networking. Int. J. Urban Reg. Res. **25**(2), 227–252 (2001)
82. B. Wellman, S.D. Berkowitz, *Social Structures: A Network Approach*, vol. 2 (CUP Archive, Cambridge, 1988)
83. D.B. West et al., *Introduction to Graph Theory*, vol. 2 (Prentice Hall, Upper Saddle River, 2001)
84. J.J. Wu, H.J. Sun, Z.Y. Gao, Cascading failures on weighted urban traffic equilibrium networks. Phys. A: Stat. Mech. Appl. **386**(1), 407–413 (2007)
85. B. Zhang, S. Horvath, A general framework for weighted gene co-expression network analysis. Stat. Appl. Genet. Mol. Biol. **4**(1), Article 17 (2005)